Fractal Antenna Design using Bio-inspired Computing Algorithms

Authored By

Balwinder S. Dhaliwal

National Institute of Technical Teachers
Training and Research Chandigarh, India
& IKG Punjab Technical University
Jalandhar, India

Suman Pattnaik

Sri Sukhmani Institute of Engineering and
Technology Dera Bassi
Punjab, India

&

Shyam Sundar Pattnaik

National Institute of Technical Teachers
Training and Research
Chandigarh, India

Fractal Antenna Design using Bio-inspired Computing Algorithms

Authors: Balwinder S. Dhaliwal, Suman Pattnaik and Shyam Sundar Pattnaik

ISBN (Online): 978-981-5136-35-7

ISBN (Print): 978-981-5136-36-4

ISBN (Paperback): 978-981-5136-37-1

© 2023, Bentham Books imprint.

Published by Bentham Science Publishers Pte. Ltd. Singapore. All Rights Reserved.

First published in 2023.

need for a court order if at any point you breach any terms of this License Agreement. In no event will any delay or failure by Bentham Science Publishers in enforcing your compliance with this License Agreement constitute a waiver of any of its rights.

3. You acknowledge that you have read this License Agreement, and agree to be bound by its terms and conditions. To the extent that any other terms and conditions presented on any website of Bentham Science Publishers conflict with, or are inconsistent with, the terms and conditions set out in this License Agreement, you acknowledge that the terms and conditions set out in this License Agreement shall prevail.

Bentham Science Publishers Pte. Ltd.
80 Robinson Road #02-00
Singapore 068898
Singapore
Email: subscriptions@benthamscience.net

BENTHAM SCIENCE

CONTENTS

FOREWORD

The book "FRACTAL ANTENNA DESIGN USING BIO-INSPIRED COMPUTING ALGORITHMS" is a Ph.D. research work based on low-cost optimized fractal antennas in order to meet the challenging requirement of compact antennas in wireless and medical applications. The objectives of this research work are mainly based on these three folds: development, optimization and experimentation of fractal antennas for low-power applications.

The book has been written in a logical and comprehensive manner with a detailed explanation of the design and methodology adopted. I am confident that this book will be of immense use to the researchers pursuing research in the domain of antenna design and optimization.

Ananta Lal Das
Society for Applied Microwave Electronics Engineering
and Research (SAMEER Ministry of Electronics and IT
Mumbai, Govt. of India

PREFACE

The demand of the compact antennas is increasing continuously due to the requirement for reduced-size wireless communication devices. The use of fractal geometry for the design of small-size antennas is a modern trend. As closed-form expressions do not exist for fractal antennas, alternative methods of designing fractal antennas are needed. The use of bio-inspired computing techniques like Artificial Neural Network (ANN), Genetic Algorithm (GA), Particle Swarm Optimization (PSO), and Bacterial Foraging Optimization (BFO) is very appropriate in such cases. In the presented research work, these techniques have been used for parameter estimation and design optimization of fractal patch antennas. Therefore, the presented research works confines the fractal antennas & bio-inspired computing techniques to provide cost-effective & efficient solutions. An extensive literature survey is carried out to understand the concept of fractal antennas, their features and design approaches. Also, a number of research papers are reviewed on the applications of bio-inspired computing techniques for antenna design, especially fractal antenna design. The extracts of the literature survey presented in the book highlight these important issues.

Many fractal antenna geometries suitable for medical and communication applications have been proposed in the presented research work. The IE3D software has been used to simulate various fractal antennas, and the simulation results are obtained to analyze the performance of the selected antennas. The desired features are assessed from the S_{11} plots, gain plots and radiation patterns which are validated with experimental and analytical findings. The multilayer perceptron neural network, radial basis function neural network, and generalized regression neural network models are developed to estimate various parameters of the proposed fractal antennas. The performance of various ANN models has also been compared in order to find optimally suitable models. The use of ANN ensemble models for the design of fractal antennas is also explored, and it has been found that the ANN ensemble approach is better than the traditional ANN model approach. The different methods of developing ANN ensemble models are also presented. The bio-inspired computing techniques based on GA, PSO and BFO are developed to find the optimal design of the proposed fractal antennas for the desired applications. The performance comparison of the various bio-inspired computing techniques is also carried out to select the best algorithm. The use of ANN models as an objective function of optimization algorithms is also enumerated to design the presented fractal antennas. It has been observed that the developed bio-inspired computing techniques provide accurate solutions with a very small computational cost. The performance of the designed antennas is validated by fabricating prototypes and then performing experimental testing. The simulated results are compared with the experimental results, and good matching of simulated and experimental results is observed in almost all cases. The obtained results are also compared with the previously published results to validate the presented designs. The research work has resulted in the design of the fractal antennas having many desirable features like size reduction characteristics, enhanced gain, and improved bandwidths

This book contains six chapters which present the outcomes of the above-described research work.

Balwinder S. Dhaliwal
National Institute of Technical Teachers
Training and Research Chandigarh
India

Suman Pattnaik
Sri Sukhmani Institute of Engineering
and Technology Dera Bassi Punjab
India

&

S. S. Pattnaik
National Institute of Technical Teachers
Training and Research Chandigarh
India

ACKNOWLEDGEMENT

The authors would like to thank IKG Punjab Technical University, Jalandhar, for providing the opportunity to publish this work.

CHAPTER 1

Recent Advances in The Design and Analysis of Fractal Antennas

Abstract: Microstrip patch antennas mainly draw attention to low-power transmitting and receiving applications. These antennas consist of a metal patch (rectangular, square, or some other shape) on a thin layer of dielectric/ferrite (called a substrate) on a ground plane. Microstrip antennas have matured considerably during the past three decades, and many of their limitations have been overcome. As the size of communication devices is decreasing day by day, the demand for miniaturized patch antennas is growing. Many methods of reducing the size of antennas have been developed in the past two decades. The recent trend in this direction is to use fractal geometry. The design of an antenna for a specific resonant frequency requires the calculation of the optimal value of various dimensions. This is a hard task for fractal antennas because the accurate mathematical formulas leading to exact solutions do not exist for the analysis and design of these antennas. The use of bio-inspired computing techniques is gaining momentum in antenna design and analysis due to rapid growth in the computational processing power, and the main techniques are Artificial Neural Network (ANN), Genetic Algorithm (GA), Particle Swarm Optimization (PSO), Bacterial Foraging Optimization (BFO), and Swine Influenza Model-based Optimization (SIMBO), *etc*. In the area of antenna design, the ANNs are employed to model the relationship between the physical and electromagnetic parameters. The trained ANNs are effectively used for the analysis and design of various types of antennas. Bio-inspired optimization techniques have been used by researchers to calculate the optimal parameters of various patch antennas and for the size optimization of antennas. Also, the hybrids of ANN and optimization techniques are proposed as effective algorithms for many applications, especially when the expressions for relating the input and output variables are not available. The presented research has addressed these recent topics by designing miniaturized fractal antennas using bio-inspired computing techniques for various low-power applications, thus, providing cost-effective and efficient solutions.

Keywords: Fractal antenna, Miniaturized antennas, Multiband antennas, Sierpinski gasket, Ultra wide band antenna.

INTRODUCTION

Antennas are used in almost all electronic devices used for wireless communication. These communications include direct person-to-person communications, communication through base station/Satellite, wireless networks like Wireless Local Area Networks (WLAN), *etc.*, and entertainment communications. The quality and efficiency of these communications largely depend on the efficient antenna design. Also, the size of communication devices is decreasing day-by-day, which dictates a very small space for fitting antennas. Therefore, miniaturized antennas are a need of the day [1, 2]. Another requirement is the design of wide-band antennas because most of the communications transfer data with complex signals composed of voice, data, images and video. The fractal antennas have the capability of miniaturized, multiband and wideband performance [3 - 6]. Also, bio-inspired optimization algorithms have the potential to provide better-quality results with reduced computation costs [7]. So, the motivation of the presented research work is to use the fractal geometry concept to provide solutions to the requirement of multiband, miniaturized and enhanced gain antennas for medical and communication applications and to use bio-inspired computing techniques to obtain the optimized fractal antennas to address the issues of antenna requirements.

ANTENNAS FOR COMMUNICATION APPLICATIONS

The antennas are the most important part of wireless communication systems. The resonant behavior of the antenna has a large effect on the communication system's performance. Most of the wireless communication applications, like Bluetooth, Wireless-Fidelity (Wi-Fi), *etc.*, work in Industrial, Scientific, and Medical (ISM) bands. The ISM bands cover frequency ranges 902-928 MHz, 2400-2484 MHz and 5725-5850 MHz, which can be used without end-user licenses. The advantage of being in the category of unlicensed bands is that there is a great scope for the development of consumer and professional products which is considered to be an important step towards the development of wireless computing, mobile internetworking, or multimedia applications. These bands have various types of applications like Bluetooth, Radio Frequency Identification (RFID), Wi-Fi, WLAN and Worldwide Interoperability for Microwave Access (WiMAX) [1]. There are two types of approaches to designing a system operating at multiple frequencies: the conventional technique using multiple single-band antennas, each intended for only one of the multiple discrete frequency bands, or a single multi-band antenna designed to handle all discrete frequency bands, *e.g.*, a fractal antenna. Another important aspect of antenna design for communication applications is to develop miniaturized antennas, *i.e.*, antennas with reduced dimensions. The miniaturization of antennas helps in designing compact wireless communication devices [2]. The bandwidth of conventional microstrip antennas is very small, so bandwidth enhancement techniques are also very essential in antenna design. There are several methods of increasing bandwidth, and the use of fractal geometry is the latest trend to achieve this [3]. The design of antennas suitable for Multi Input Multi Output (MIMO) systems is also attracting the attention of antenna designers because this technology enhances the data transmission capacity and reduces multipath fading effects [4]. The design of wearable antennas, which are flexible enough that these can be bent, crumpled, and folded, is also another recent trend. These antennas are generally stitched as part of clothes and are used for many applications such as military, health monitoring activities, telemedicine, sports, tracking, *etc* [5]. The main challenge in designing wearable antennas is to find appropriate fabrics and polymers which can be employed as flexible substrate materials. The other important challenges are the design of antennas having high gain [6] and circular polarization [7].

ANTENNAS FOR MEDICAL APPLICATIONS

Antennas for medical applications have been widely investigated and reported in the recent past. The recent applications are typically in the field of information transmission, such as RFID / wearable or implantable antennas, in diagnoses such as Magnetic Resonance Imaging and microwave computed tomography/ radiometry, and also wireless telemedicine / mobile health systems. Applications are also reported in thermal therapy (hyperthermia, coagulation, *etc.*) and microwave knife [8]. Most modern Implantable Medical Devices (IMD) help in establishing a communication link between the implant and external devices behaving as a telemetry system. This communication link can be used to temporarily or permanently program the operating parameters of the IMD, to retrieve both real-time and stored physiological data, and to enquire about the IMD system status and therapy history. Several techniques aiming at creating a physical channel have been developed for IMD telemetry, namely static magnetic field coupling, reflected impedance coupling and Radio-Frequency (RF) propagation. Recently, RF transmissions have received increased attention because of their higher data rates and ability to communicate over long distances between the IMD and the external device [9]. For medical data telemetry, the Medical Implant Communication Service (MICS) band (402-405 MHz) was established by the Federal Communications Commission (FCC) in 1999, and the ISM frequency bands are also available. However, most of the transceivers make use of the MICS and 2400 MHz ISM bands [10]. Hence, by providing communication of the sensor with external equipment, antennas find a major role in medical systems. Small size and high radiation efficiency are the main challenges faced by antenna designers for medical applications. Other than these, some other issues like impedance matching, low-power requirements, and biocompatibility with the body's physiology, directivity, lobe control, *etc.*, are also considered while designing antennas [11, 12].

LIMITATIONS OF EXISTING ANTENNA SYSTEMS

The limitations of existing antennas used in medical and communication applications are as follows:

- Moderate gain.
- Limited directivity.
- Large size.
- Limited bandwidth.
- Not very suitable for MIMO applications due to lack of reconfigurability.
- Non-availability of accurate & efficient tools for fractal antenna design.
- Hybridization of fractal geometries with other techniques has not been investigated.

FRACTAL ANTENNAS

The use of fractal geometry for the design of small-size patch antennas is a recent development in the direction of size reduction and multi-band performance. The definition of 'Fractal' was given by Benoit Mandelbrot in 1975. According to Mandelbrot, fractal geometry is a way of classifying structures whose dimensions are fractional numbers [13]. The fractal geometries are uneven shapes which can be separated into sub-parts, and every sub-part is (at least approximately) a small copy of the overall shape. Examples of mathematical fractal geometries are Sierpinski's gasket, Von Koch's snowflake, Cantor's comb, the Lorenz attractor, the Mandelbrot set, *etc*. The real-world examples of fractal shapes include mountains, clouds, turbulences, and coastlines, which cannot be represented by Euclidian shapes [14]. Fig. (**3.1**) of Chapter 3 shows a fractal geometry named Sierpinski gasket fractal geometry.

The antennas which use fractal geometry as radiating structures are known as fractal antennas. These antennas use self-similar and space-filling properties of the fractal geometries to design antennas which have more electrical length fitted into a small area. The fractal antennas are of small size and therefore expected to have many important applications in wireless communication. The advantages of fractal antennas [15] include:

- Miniaturization and space-filling
- Multiband performance
- Efficiency and effectiveness
- Improved directivity
- Improved gain

DESIGN AND ANALYSIS OF FRACTAL ANTENNAS: RECENT DEVE-LOPMENT

A literature survey on fractal antennas has been carried out by referring to many National & International journals and conference proceedings such as the Journals and Transactions of IEEE (Antennas and Propagation, Antenna and Wireless Propagation, Microwave Theory and Techniques *etc.*), Journal of Progress in Electromagnetics Research, Microwave and Optical Technology Letters, Journal of IETE, Proceeding of various International and National conferences, and various books. The extracts of the most pertinent observations are:

Puente *et al.* [16] introduced the Sierpinski gasket-based monopole fractal antenna having multiband performance. The radiating structure of the antenna is made on a dielectric substrate and fixed perpendicularly on a ground plane. The experimental and computed results are presented to show a multiband behavior over five bands for this fractal Sierpinski antenna. This behaviour is due to the self-similarity characteristics of the fractal geometry of antenna.

Puente-Baliarda *et al.* [17] discussed in detail the operation of a multiband fractal antenna based on Sierpinski triangle. The described fractal antenna has shown a notable degree of similarity at five bands, the same number of scales over which the fractal structure appears similar. The bands are also spaced by a log period of two, the same spacing that relates the five scales on the fractal shape. Thus, it is concluded that the geometrical self-similarity properties of the fractal structure have been translated into its electromagnetic behaviour. Due to its mainly triangular shape, the antenna is compared to the well-known single-band bow-tie antenna.

Werner *et al.* [18] described the theory and basics needed for analyzing and designing the arrays of fractal antennas. They also introduced various types of fractal arrays and outlined many essential features of these arrays, which include the multi-band behaviour, methods of implementing structures with small side-lobe levels, efficient designs to thinning, and the capability to design speedy beam-forming algorithms by using the features of fractal geometry.

Best [19] analyzed the radiation patterns of the Sierpinski gasket fractal antenna and found that the self-similar features of return loss characteristics are not observed in radiation pattern characteristics. He proposed a modified Parany gasket antenna and established that the radiation pattern features of this antenna are quite similar to the Sierpinski gasket antenna.

Gianvittorio and Rahmat-Samii [20] outlined the size reduction properties of fractal geometries and proposed its use for designing wire and patch antennas. They also described that although mathematically fractal geometry has infinite iterations, but for fractal antennas, the first few iterations are sufficient. They also used the compact fractal antennas to develop the phased arrays.

Tang and Wahid [13] proposed a fractal antenna using the hexagon as the base shape and found that multi-band characteristics can be achieved using this shape. The first three iterations are analyzed and it is found that the resonant frequencies of adjacent bands have ratios equal to three, which is two in the case of the Sierpinski triangular fractal antenna. This large value of resonant frequency isolation results in the increased advantage of flexibility in implementing multi-band applications.

Best [21] considered different wired fractal monopole antennas and compared their resonant properties. The fractal geometries analyzed are Hilbert, Minkowski, Koch, Tee, and some Meander-line geometries. The efficiency, radiation resistance, and quality factor of antennas are the parameters which are evaluated. It is established that these antennas have similar behavior for the same area and the same wire diameter.

Werner and Ganguly [14] presented a comprehensive overview of the research in the area of fractal antenna engineering. The various topics considered are the design methodologies for fractal antenna elements, the application of fractals to the design of antenna arrays, and frequency-selective surfaces with fractal screen elements. The Iterated Function System (IFS) used for designing fractal shapes is described. The hybrid of GA and IFS is also explained for designing the fractal wire antennas.

Dehkhoda and Tavakoli [22] described a crown square microstrip fractal antenna and demonstrated that the size is reduced compared to a nearly square antenna at the first resonant frequency. At higher resonant frequencies, the crown square antenna has a larger Voltage Standing Wave Ratio (VSWR) bandwidth. The antenna is considered up to the second iteration, as the further iterations do not have much effect on the resonant properties.

Anguera *et al.* [23] presented a multiband fractal antenna designed using the modified Sierpinski fractal shape and two parasitic patches. Multilayer arrangement is used to implement the antenna showing dual band behavior and broad bandwidth. The radiation pattern shapes are almost similar in both bands. An electrical circuit model is proposed to demonstrate the enhancement of input impedance.

Rahim *et al.* [24] described a square patch fractal antenna based on Sierpinski carpet geometry. The antenna is excited using the transmission line feeding technique. The return loss and the radiation pattern parameters are used to study the behavior of the antenna. The multiband operation is observed from simulation and experimental results. The radiation pattern has shown that the proposed fractal geometry has a performance comparable to a dipole antenna.

Ding *et al.* [25] presented a crown fractal antenna based on a circular shape. The antenna is developed up to the second iteration and fed using the Co-Planar Waveguide (CPW) feed. The resonant parameters of the antenna show that the antenna has ultra-wideband (UWB) performance with wideband and omni-directional features. The antenna has very small size dimensions.

Huang *et al.* [26] designed a fractal antenna with a tuning stub. The antenna has multiband and broadband behavior. A microstrip line is used to excite the antenna. The proposed antenna has a reduced size and simple structure. A bandwidth of almost 18% is achieved using the proposed fractal geometry, which is a high value for single microstrip antennas.

Anguera *et al.* [27] proposed a triple-band fractal antenna implemented using the combination of a double-band and a single-band antenna. Multilayer stacking is used to develop the antenna. All three bands have high efficiency and broad bandwidths. The antenna has similar radiation pattern shapes in different bands.

Song *et al.* [15] demonstrated that the Sierpinski gasket geometry-based fractal patch antennas can be designed to operate at desired frequencies by varying the fractal designs and iterations. They designed a double band fractal antenna with a modified Sierpinski gasket shape having a broadband behavior. The perturbation factor of the antenna is varied to get the improved bandwidth. The results of the modified Sierpinski antenna are compared with the standard Sierpinski antenna.

Guangguo and Shouzheng [28] proposed two fractal monopole antennas designed using the hybrid of Koch and Sierpinski gasket fractal geometries. The fractal elements miniaturize the traditional Sierpinski gasket antenna while maintaining the multi-band behavior. Micro-electromechanical System (MEMS) switches are used to introduce the frequency reconfigurable characteristics, which results in increased flexibility in multi-band operation.

A reduced-size multiband microstrip antenna is presented by [29]. The square antenna is taken as a base antenna, and it has slots based on Sierpinski fractal geometry. The antenna has a multilayer structure and is excited electromagnetically by a microstrip line printed on other substrates. The antenna has a double resonant band and a size miniaturization of around 77% at a fundamental resonant frequency.

Hwang [30] used a vertically cut half Sierpinski triangle to implement an antenna with multiple resonant bands. The half Sierpinski triangle is further modified and the slotted ground plane is utilized to attain the broadband characteristics. The antenna is excited using a co-axial probe feeding technique and has a nearly omni-directional pattern.

Kumar *et al.* [31] presented a rectangular antenna with two sides having fractal geometry. The antenna is implemented using a single substrate with co-axial feeding. The multilayer configuration is also designed, and proximity coupling is employed with and without air-gap. The multilayer antenna results in improved bandwidth and reduced size. The proposed multilayered fractal antenna has a better cross-polarization level which varies with the size of the air gap.

Azari and Rowhani [32] presented new fractal geometry for microstrip antennas based on hexagonal shapes. The antenna is developed up to the second iteration and has UWB operations. Co-axial feeding is used for exciting the antenna. The radiation pattern plots show that the antenna has a good value of gain.

Naghshvarian-Jahromi [33] introduced a fractal patch antenna using pentagon-shaped Koch geometry. The antenna is fed using microstrip line feed, and the performance of the planar structure is compared with the monopole antenna performance. The antenna is also analyzed in the time domain. The antenna has sufficient gain for the WLAN and Bluetooth applications.

Mishra *et al.* [34] developed an updated expression for Sierpinski gasket monopole fractal (SGMF) antenna. The expression determines the resonant frequency of the antenna for given values of fractal iteration number, the length of the outermost triangular shape, and substrate parameters. The expression for the design of an antenna for the given value of frequency is also proposed.

Yong and Shaobin [35] proposed a fractal antenna based on the square base shape having multiple resonant bands. A method to control the frequency separations between different bands is also proposed. The microstrip line feeding method is used to improve the matching properties. The antenna covers the Global Positioning System (GPS) and digital multimedia broadcasting applications.

Mirzapour and Hassani [36] proposed a fractal antenna based on a snowflake shape. The antenna has a compact size and wide bandwidth of operation. The performance of the antenna with a coaxial probe and capacitively coupled feeding are compared. The bandwidth improvement of about 50% is achieved by using an air-filled substrate and capacitive feed. The miniaturization of almost 70% is attained by the slot-loading technique as compared to a simple Koch fractal antenna.

Krzysztofik [37] presented a fractal monopole antenna for ISM band applications using modified Sierpinski fractal geometry. Considerable miniaturization is attained in comparison to the standard Sierpinski triangular antenna by varying the separation of the first two resonant frequencies. The use of the proposed ante-

-nna for handset applications is explored by studying the effect of hand and head on the performance of the antenna.

A fractal-shaped slot is used for enhancing the bandwidth of a square patch antenna [3]. A microstrip line is used for feeding the antenna. The dependence of the bandwidth on the iteration order and iteration factor of the fractal geometry is analyzed. The bandwidth improvement results are compared with a traditional square antenna with wide slot and an improvement of around 3.5 times is achieved.

Manimegalai *et al*. [38] presented a fractal monopolar antenna based on Cantor set geometry. The antenna has multiple resonant bands and covers many applications in 2 GHz to 7 GHz frequency band. The Cantor set geometry helps in size reduction and provides flexibility for controlling the resonant frequencies and their bandwidths.

Cao *et al*. [39] presented a fractal antenna based on Minkowski geometry. The antenna is implemented for RFID reader applications in 2.4 GHz band, and it has a bandwidth of 100 MHz. The dimensions of the proposed fractal antenna are lesser than the traditional patch antenna based on a rectangular shape. The small size is helpful in designing miniaturized RFID systems.

Gemio *et al*. [40] employed the fractal geometry-based ground plane to design a multiband antenna. The triangular patch is printed on a substrate and fixed over a fractal ground plane in a perpendicular direction to design a monopole configuration. The combination of resonances of fractal ground and that of monopolar patch enhance the features of the entire geometry. The antenna is designed to operate over two resonant bands covering WLAN applications of 802.11 standard.

Aggarwal and Kartikeyan [41] presented the design of a fractal patch antenna, which uses a unique fractal geometry known as Pythagoras tree. The antenna is excited using CPW feed and it has been designed for double band resonant behavior at the 2.4 GHz and 3.5 GHz band for a very wide bandwidth feature.

Aggarwal and Kartikeyan [42] proposed, simulated and tested a fractal patch antenna using modified Sierpinski geometry. The antenna has multiple resonant bands. The wideband features are attained using a slotted ground plane and the modified Sierpinski structure.

Anoop *et al*. [43] proposed a patch antenna with fractal geometry having multiple resonant bands. The antenna has a compact size, less weight and the fractal geometry resulted in multiband characteristics. The base shape of the fractal antenna is square and the first three iterations are explored to analyze the performance.

Jibrael and Hammed [44] designed a fractal patch antenna with plus-shaped fractal geometry. The proposed antenna has a small size and multiple frequency operations. The self-similarity and space-filling characteristics are observed to a high level. The designed antenna has four resonant bands between 2 GHz and 12 GHz.

Azari [45] has presented a fractal antenna using an octagonal geometry. The proposed antenna is explored up to the second iteration and has super wideband features making antenna suitable over a wide frequency range from 10 GHz to 50 GHz. The antenna has a good value of maximum gain over the whole resonant band.

Bayatmaku *et al.* [46] presented a patch antenna using an E-shaped fractal geometry. The antenna is excited using a co-axial feed. The different iterations and various combinations of the parameters in every iteration are explored to find an optimal design suitable for Long Term Evolution (LTE) / Wireless Wide Area Network (WWAN) applications. The antenna performance is analyzed using various parameters like bandwidth, gain, radiation patterns, and return loss.

Choukiker *et al.* [47] presented an antenna employing fractal geometry in half-circle form derived using the Descartes circle theorem. The antenna is excited using a modified CPW feed. The antenna has a double band of resonance covering 2.4/5.2 GHz WLAN applications with sufficient bandwidth for these bands. The current distribution characteristics of the antenna are analyzed to understand the resonant behavior.

Soh *et al.* [48] presented the use of Sierpinski geometry for the design of a planar inverted-F antenna. The antenna is designed using textile and polyester fabrics, so it is suitable for on-body integration. The antenna has a dual frequency of operation centered at 2.45 GHz and 5.2 GHz with good bandwidth. The shape of the radiation pattern is omni-directional with a moderate value of gain. The antenna has an efficiency of around 70%.

Sung [49] has proposed a wideband fractal slot antenna using Sierpinski fractal geometry. The base shape is a square patch antenna with a wide square slot. The square slot is modified by the Sierpinski structure, keeping the same outer dimensions. The antenna is excited using a microstrip line feed. The number of resonant bands depends upon the iterations of the Sierpinski square radiating elements. The fractal slot structure also enhances the bandwidth of the slot antenna.

Lee *et al.* [50] investigated the design of a wearable Minkowski fractal antenna for Very High Frequency (VHF) band. Two different conductive materials are examined to check the suitability for the desired band. An L-shaped folded ground plane is used while feeding the antenna. The proposed flexible antenna achieved an efficiency of 48%, with a size of less than 0.5 m.

Pourahmadazar *et al.* [51] presented a fractal monopole antenna using Pythagoras tree geometry. The antenna is excited using a modified microstrip line feed and UWB performance. The traditional T-patch is modified by inserting a Pythagoras tree geometry to achieve a wide bandwidth. The proposed antenna has a reduced size, and the proposed multi-fractal concept results in flexibility for shifting resonances and bandwidth.

Jahromi *et al.* [52] enhanced bandwidth and impedance matching of fractal monopole antennas by using the grounded CPW feeding technique. The results of this new feeding method are evaluated against the conventional standing monopole antenna and a significant improvement in magnitude of S_{11} parameters is observed. It is found that the grounded CPW feed results in antennas which show a small cross-polar field and a good radiation pattern.

Kumar and Nikam [53] presented a fractal antenna using modified Appollian gasket fractal geometry. The antenna is fed using CPW. The fractal geometry in combination with a modified ground plane, resulted in UWB operation. The effect of variation of different parameters of antenna is studied. The antenna shows a bandwidth of almost 143% centered at 10.5 GHz with a nearly omni-directional radiation pattern.

Oraizi and Hedayati [54] used square and Giuseppe Peano fractal shapes to design a microstrip antenna with multiple frequency operation. The antenna is implemented as a double layer structure with radiating patch on the top and ground plane at the bottom. A microstrip line is used in between the two substrate layers to feed the antenna by electromagnetic coupling. The antenna exhibits a circularly polarized behavior.

Kumar *et al.* [55] presented a monopole fractal antenna based on inscribed triangle circular geometry. The antenna is excited using the CPW feed and has UWB operation. The radiation pattern of the antenna has omni-directional shape. The parametric study of the antenna is presented to show the effect of different parameters on the antenna behaviour.

Oraizi and Hedayati [56] also investigated the use of Giuseppe Peano fractal shapes for designing compact microstrip antennas. The boundary of the square patch is modified in the form of Giuseppe Peano curves which result in increased perimeter in almost same surface area which results in size reduction while maintaining the gain. Slotting of the radiating element produces additional size reduction. The broad-band characteristics are obtained by inserting an air-gap between the radiating element and ground plane.

Olaode *et al*. [57] presented the approach of Meandering *i.e.*, inserting bends for achieving the miniaturization. They proposed analysis of the results of replacing the straight dipole antenna by the Meander fractal curves. The dipole antenna operating in VHF range is selected for the verification. The equivalent circuit is also derived from that of straight-line dipole. The meandered dipole antenna up to 6 bends is analyzed and it is found that the antenna with three bends has optimum performance.

Li and Mao [58] modified standard bow-tie antenna by using the Koch-like fractal geometry. The sides of the bow-tie antenna are replaced with fractal curve and the performance is compared with the unmodified bow-tie dipole and triangular Sierpinski antenna. It is found that all the antennas have the same resonant bands in low frequency but improved resonant parameters in high-frequency bands.

Ghatak *et al*. [59] presented a Sierpinski fractal antenna with hexagonal boundary. The antenna is analyzed up to second order and UWB performance is achieved with band rejection features. A 'Y' shaped slot is inserted to achieve the band rejection characteristics. The antenna is excited using CPW feeding mechanism. The time domain analysis shows that the group delay is very small over whole band.

A Sierpinski fractal antenna for THz band is designed by Xu *et al*. [60]. The electrical conductivity of the graphene varies with the frequency and this property is used to make the antenna reconfigurable. The slots of Sierpinski geometry are filled by the graphene to achieve the reconfigurable characteristics. The S_{11} and radiation patterns plots of first three iterations are explored to study the behavior of the antenna.

Dorostkar *et al*. [61] presented a fractal antenna based on circular and hexagonal shapes. The antenna has a super wideband operation, which is achieved by a partial ground plane. GA is used to optimize the parameters. The antenna has good impedance bandwidth and gain, which makes it attractive to many wireless applications.

Kumar and Choubey [62] presented a circular pentagonal fractal antenna for UWB operation with a notch band to escape obstruction from the WLAN communications. The CPW feeding is used to excite the proposed antenna. A ring slot on the feeding line is used to insert the notch band at desired frequency. The antenna has omni-directional shape of radiation pattern in H-plane and bidirectional shape in E-plane.

Naser-Moghadasi *et al.* [63] proposed a CPW fed monopolar antenna with a fractal shaped radiating structure and a T-shaped structure inbuilt into it. The antenna has UWB operation with a notch band between 3.3-4.2 GHz. The UWB characteristics are obtained by using rectangular notches in the ground plane and the notch features are obtained by the T-shaped structure. A bandwidth of around 117% is achieved, excluding the notch band.

Two triple band antennas are presented by the Varadhan *et al.* [64] using tree like fractal geometries. One antenna is designed for RFID reader and other is for RFID tag. A read range up to 87.5 cm is achieved by the tag antenna. The antennas have sufficient bandwidths for these applications. The CPW feed is used for exciting the antennas.

Karmakar *et al.* [65] presented a monopolar antenna having fractal shaped slots. The antenna has UWB performance which is achieved by slotting the ground plane and impedance steps. The impedance matching is improved by the adjusting the gap between the radiating patch and the ground plane. The antenna has a bandwidth of around 120% with an omni-directional radiation pattern with sufficient gain for UWB applications.

A compact fractal antenna using the modified rectangular Sierpinski geometry is proposed by Shrestha *et al.* [66]. The antenna is designed for ISM band applications. The inset feeding is used to excite the patch and it resulted in good impedance matching. The antenna is developed up to second iteration Sierpinski carpet geometry. A miniaturization of around 32% is achieved with reasonable values of return loss and gain of antenna.

Thi *et al.* [67] presented a planar Spidron fractal antenna for Ku-band satellite communication. The antenna has double band of resonant frequencies with circular polarization. The antenna is excited using the microstrip line feeding and impedances bandwidth up to 9% is achieved. The 3 dB axial ratio bandwidth of around 3% is obtained for the first band. The proposed antenna has a small operating frequency ratio of 1.15 which is required for satellite communications.

Fallahi and Atlasbaf [68] presented a monopole antenna having UWB performance. The small size fractal shapes are added to the corners of the base polygon radiating structure. The CPW feeding is used to excite the bandwidth and the antenna has a compact size. The antenna is analyzed in time domain by calculating the fidelity factor which is more than 0.92. This means that the antenna is suitable for radar applications.

Pakkathillam *et al.* [69] presented a tapered slot antenna in which the slot boundary is developed using a fractal shape. The antenna is developed for ultra-high frequency RFID reader applications using the variable distance opposite direction current concept. The antenna has a bandwidth of almost 101 MHz centered at 897 MHz.

A multiband antenna for RFID reader applications is proposed by Liu *et al.* [70] in their paper entitled "Dual-Band Microstrip RFID Antenna with Tree-Like Fractal Structure". The tree-like fractal structure is placed in between the patch and ground plane of an air-filled rectangular microstrip patch antenna to achieve dual-band operation. The two resonant bands have impedance bandwidths of almost 4.4% and 3.1%.

Liu *et al.* [71] in their paper entitled "Miniaturised Wideband Circularly-Polarised Log-Periodic Koch Fractal Antenna" presented a compact fractal antenna with Koch log-periodic shape. The antenna has a resonant band of 2 GHz to 6 GHz with circular polarization. Two crossed dipoles are employed to achieve circularly polarized behavior and the broadband features are attained by a log-periodic dipole antenna.

Choukiker and Behera [72] presented a small-size sectoral fractal planar monopole antenna. The shape of the antenna is inspired by the Sierpinski gasket shape. The antenna has a wideband of operation covering WiMAX, Wi-Fi and WLAN applications. The antenna is excited by the microstrip line feed through a matching network.

Dholakiya and Pujara [73] used a circular-shaped fractal slot to design an enhanced bandwidth antenna. The antenna performance is analyzed for different iteration factors and orders. A microstrip line feed is used to excite the antenna and it is designed on FR4 substrate. A bandwidth of around 53% is achieved with reasonable value of gain.

A triple band wearable antenna is presented by Jalil *et al.* [74] using Koch fractal geometry. The simulation model is developed using the CST software. The denim textile material is used as substrate and two different conducting materials are used to fabricate two different antennas. The antenna performance is studied under three different conditions. The S_{11} and bandwidth parameters are taken as the performance evaluators.

Xu *et al.* [75] used meta-surfaces and meta-resonators to design miniaturized patch antennas. The antennas have circular polarization, which is achieved by employing meta-resonators designed using crossbar fractal tree slot and spiral resonators. The miniaturization is attained by the use of metamaterial reactive impedance surface. The proposed concept is verified by implementing three different antennas.

Wang *et al.* [76] proposed a fractal antenna based on modified Sierpinski-carpet geometry. The fractal geometry is developed up to third iteration. The antenna has triple resonant band characteristics between 1 GHz to 20 GHz. The lower band covers whole UWB frequency range. The antenna has good radiation pattern with sufficient gain in resonant bands.

[6] presented a fractal antenna for X band and Ku band applications. The traditional rectangular patch antenna is modified by adding fractal geometry at the corners and sides. The introduction of fractal shapes resulted in improvement in gain and bandwidth of the antenna. The designed antenna has multiband operation and has linear polarization. The microstrip line feeding is used for the excitation.

[4] presented the design of a fractal monopole antenna for handheld devices and suitable for MIMO implementation. The antenna shape is a hybrid fractal geometry designed by merging the Minkowski and Koch fractal shapes. The antenna has dual band performance with good bandwidth in both bands.

An antenna with fractal slots is proposed by Zarrabi *et al.* [77] for double notch application. The antenna is designed on FR4 substrate and has UWB performance. It operates over 2.1-12 GHz band with two notch frequencies at 3 and 5 GHz. The designed antenna has sufficient gain for wireless applications in this frequency band.

Jalali and Sedghi [78] proposed a small size CPW fed monopole fractal antenna having UWB performance. Slots are created in the ground plane to achieve the UWB performance. The resonant band for the antenna is from 2.95 GHz to 12.81 GHz resulting in impedance bandwidth of around 125%. Also, the antenna has omni-directional radiation pattern over whole resonant frequency band.

Choi *et al.* [79] designed a miniaturized fractal antenna for rectenna system working at a frequency of 2.45 GHz. The rectangular Sierpinski carpet structure is used to reduce the size of the antenna. The RF energy is harvested by the fractal antenna from the environment and the output of the antenna is applied as input to a rectifier circuit to produce the dc output.

A circularly polarized patch antenna is designed by Reddy and Sarma [80] using the fractal boundary approach. The straight boundary of a square patch is replaced by the asymmetrical pre-fractal curves which lead to the excitation of two perpendicular modes resulting in circular polarization. The fractal sides also result in size reduction of the antenna.

Reddy and Sarma [81] also presented a triple band Koch fractal boundary and fractal slot microstrip antenna having circular polarization. The four different versions of the antenna are explored and the final antenna is proposed to be suitable for WLAN/WiMAX wireless applications. The antenna has good 3-dB axial-ratio bandwidth which is another advantage for modern wireless applications.

A quasi-self-complementary microstrip patch antenna having UWB character-istics is proposed by El-Hameed *et al.* [82]. The antenna geometry employs crossbar fractal boundary for achieving miniaturization. The impedance matching is improved by inserting a slot in ground plane and its step-tapering is done to reduce the lower-side value of the resonant frequency band. It has an omni-directional radiation pattern with sufficient gain for UWB applications.

Subramaniam *et al.* [83] employed Minkowski fractal curve to design flexible antennas. The Minkowski geometry results in miniaturization. The antennas are designed for WLAN applications and have acceptable gain values for these bands. Two different materials Flectron and Zeit are analyzed to implement the wearable antennas and it has been concluded that the Zeit antenna has relatively better results.

Li *et al.* [84] integrated quasi-Sierpinski fractal dipoles and high-impedance surface to design a reconfigurable fractal antenna. The frequency of the proposed antenna can be shifted between X-band, Ku-band, and Ka-band using two switches. The high-impedance surface is implemented by arranging square patches with square slots in the form of an array and is used as a reflector to attain the improved front-to-back radiation.

Raviteja *et al.* [85] proposed a CPW-fed fractal monopole antenna for RFID reader applications at 900 MHz. The antenna has a circular polarization for a bandwidth of 36 MHz; however, the −10 dB bandwidth of the antenna is 256 MHz. The antenna has a compact design and has a read distance of 1.32 m. Therefore, the proposed antenna is suitable for short-distance RFID reader applications.

Kumar *et al.* [86] presented a fractal antenna having resonant frequency varying capability. The modified Sierpinski triangle monopole antenna has been designed using the moving feeding technique to obtain different resonant frequencies. A microstrip feed line is used as the main feed and a movable coaxial feed is attached to it. A programmable motorized method is used for experimental verification of the proposed approach.

Weng and Hung [87] proposed an H-fractal having multiband performance. The antenna is fabricated using FR4 substrate and has good directivity. The antenna has non-overlapping structure and can be designed to work at different desired bands. The design of this H-fractal antenna is achieved for 2.4/5.5-GHz WLAN bands with sufficient bandwidth for these applications.

A plus-based fractal slotted array has been implemented for X-band applications by Chatterjee *et al.* [88]. An asymmetric iris is used to excite the slots placed in the centre of the broad wall of a waveguide. The empirical formula has been derived for the conductance of the fractal unit cell. The parametric study has been done, and it is found that the iris width does not affect the array element if its width is more than that of slot.

Zhou *et al.* [89] presented a tree fractal-based antenna array which is fed by an L-shaped microstrip line. Two elements have been used to implement a 1 x 2 array which are coupled by a slot. The antenna element resonates from 1.11 to 1.71 GHz with circular polarization, which is maintained in the array configuration also. The array results in an improved value of gain.

A miniaturized reflectarray element is presented by Costanzo and Venneri [90] using the Minkowski fractal geometry. The patch dimension is kept unchanged and scaling factor of the fractal is utilized to attain the correct phase tuning. A miniaturization of approximately 30% is attained for achieved X-band operation. The designed reflectarray element has wide-angle scanning potential and can be used for broadband applications.

Authors of [5] proposed a wearable fractal monopole antenna having dual-band performance. A foam sheet of 2 mm thickness is used as the substrate and an electromagnetic band-gap structure is incorporated into the antenna to reduce the effect of human body. The behavior of antenna under different conditions is presented and specific absorption rate is also measured to validate the performance.

Tripathi *et al.* [91] presented the design of a compact fractal MIMO antenna having UWB performance. The Koch fractal geometry is used for designing the monopole elements, which are then placed in orthogonal configurations. Isolation is improved by the use of grounded stubs and WLAN band rejection is attained by

C-shaped slots. The proposed antenna has a quasi omni-directional radiation pattern.

Tripathi *et al*. [92] also proposed a small size UWB fractal antenna. The Sierpinski geometry is employed to achieve compactness and broad band behavior. The designed antenna has a radiation pattern with omni-directional characteristics, a constant fidelity factor, and very less return loss.

A compact fractal antenna with UWB performance is proposed by Singhal *et al*. [93]. The antenna employs an inner tapered tree-shaped geometry with CPW feed which is used to enhance the bandwidth. The designed antenna operates over the frequency range from 4.3 GHz to 15.5 GHz with a nearly omni-directional pattern.

Pakkathillam and Kanagasabai [94] achieved broadband characteristics by using fractal geometry in the form of slots placed along the diagonal of the patch antenna. The proposed antenna has a circular polarization, high gain, compact size, and it is suitable for handheld devices. The antenna performance is studied for different iteration factors and iteration orders.

A Spidron fractal dielectric resonator antenna having circular polarization is proposed by Altaf *et al*. [95]. A C-shaped slot is used to achieve a broad 3 dB axial ratio bandwidth. The coupling between the slot and microstrip line is used to excite the antenna. The antenna has gain values between 2.2 dBi to 3.16 dBi for the axial ratio bandwidth.

Cai *et al*. [96] presented a small-size antenna designed using fractal metasurface and fractal resonator. A reactive impedance surface based on the Hilbert fractal geometry has been employed to improve the antenna characteristics. The miniaturization and circular polarization are achieved using a ring resonator with splits designed using Wunderlich fractal geometry. The antenna performance is validated experimentally for WiMAX band applications.

A square antenna employing log-periodic fractal geometry is proposed by Amini *et al*. [97] for UWB applications. Two types of squares: squares with ring slot and square without slots, are arranged in log-periodic form to achieve the UWB operation. This geometry has resulted in a size reduction of 23 percent with almost constant gain in whole band. The antenna features in the time domain are also studied and the experimental measurements are used to validate the simulation results.

A paper substrate has been employed to develop a small size flexible UWB antenna by [98]. The antenna shape has been designed using fractal geometry concept using a combination of circular slot and octagonal patch to enhance bandwidth and achieve size reduction. The environment-friendly substrate and low-cost fabrication make this antenna suitable for many applications like WiMAX, Wi-Fi, RFID, *etc.*

Mokhtari-Koushyar *et al.* [99] proposed a flexible antenna having tree-shaped geometry inspired by fractal geometry concepts. The low-cost fabrication has been achieved using the inkjet printing technique on a flexible substrate. The CPW feed has been used, and six iterations of the antenna geometry have analyzed using simulation and fabricated results.

An innovative wideband polymer and fabric fractal monopole antenna is presented by [100]. A conductive fabric thin sheet has been employed for the conductive parts, and a composite made of natural rubber has been used for the nonconductive parts resulting in a thin and flexible prototype. Simulated and measured results According to the findings, the antenna covers the most widely used standards in wireless communication systems with acceptable radiation patterns and gain.

Kirtania *et al.* [101] outline that the demand for flexible antennas is due to many new applications of wearable devices and 5G technology. They presented an extensive review of flexible antennas designed using fractal and non-fractal antenna shapes. The other topics emphasized in the review article are materials and processes used for antenna prototyping.

A review of the printed wearable antennas developed on flexible substrates has been presented by [102], highlighting the fundamentals, characteristics, and development challenges. It was pointed out that the fabrics have very low dielectric constants because air is present in the gaps between them, so the characteristics of the flexible substrate materials are also the subject of a comprehensive analysis of antenna performance. The latest advancements in wearable antennas for use in medical applications are discussed.

A small-size wearable antenna on textile substrate has been presented in [103] for 2.45 GHz band applications. The size reduction has been achieved with the use of fractal geometry named as crown rectangular fractal antenna. The comparative analysis of five different flexible textile materials has been performed to select the best suitable substrate, and it has been proposed that denim substrate-based antenna requires relatively the least area for the same frequency. The experimental bending analysis has also been performed to determine the safe bending radius limits.

CONCLUSION

The chapter starts with a note on the foundations of miniaturized microstrip antenna design. The underlying principle of bio-inspired computing techniques for designing antenna parameters is highlighted. The features of antennas for medical and communication applications are described. The meaning of fractal antennas and their advantages are discussed. The extracts of important papers on fractal antennas published in the last two decades are provided, which highlight that fractal antennas are very suitable for designing miniaturized and multiband antennas. It is observed from the literature survey that the Sierpinski gasket is the most popular geometry used for designing fractal antennas. Different modified forms of the standard Sierpinski antenna are also proposed, which have better performance than the standard shape in one or more aspects. A number of other geometries are also proposed for fractal antenna design. The co-axial feed is used by most of the researcher; however, a good number of papers are also published which employed other forms of excitation. The CPW feed is mainly used for designing UWB fractal antennas. The use of fractal-shaped slots for enhancing the performance of microstrip antennas is also reported. The design of hybrid fractal antennas, *i.e.*, fractal antennas developed using the combinations of two fractal geometries, is presented in a few papers. The suitability of fractal antennas for various applications is explored by different researches, and is discussed in this chapter.

REFERENCES

[1]   D.D. Krishna, M. Gopikrishna, C.K. Anandan, P. Mohanan, and K. Vasudevan, "CPW-fed Koch fractal slot antenna for WLAN/WiMAX applications", *IEEE Antennas Wirel. Propag. Lett.,* vol. 7, pp. 389-392, 2008.
[http://dx.doi.org/10.1109/LAWP.2008.2000814]

[2]   H.Y.D. Yang, "Miniaturized printed wire antenna for wireless communications", *IEEE Antennas Wirel. Propag. Lett.,* vol. 4, pp. 358-361, 2005.
[http://dx.doi.org/10.1109/LAWP.2005.857033]

[3]   W.-L. Chen, G.-M. Wang, and C.-X. Zhang, "Bandwidth enhancement of a microstrip-line-fed printed wide-slot antenna with a fractal-shaped slot", *IEEE Trans. Antenn. Propag.,* vol. 57, no. 7, pp. 2176-2179, 2009.
[http://dx.doi.org/10.1109/TAP.2009.2021974]

[4]   Y.K. Choukiker, S.K. Sharma, and S.K. Behera, "Hybrid fractal shape planar monopole antenna covering multiband wireless communications with MIMO implementation for handheld mobile devices", *IEEE Trans. Antenn. Propag.,* vol. 62, no. 3, pp. 1483-1488, 2014.
[http://dx.doi.org/10.1109/TAP.2013.2295213]

[5]   S. Velan, E.F. Sundarsingh, M. Kanagasabai, A.K. Sarma, C. Raviteja, R. Sivasamy, and J.K. Pakkathillam, "Dual-band EBG integrated monopole antenna deploying fractal geometry for wearable applications", *IEEE Antennas Wirel. Propag. Lett.,* vol. 14, pp. 249-252, 2015.
[http://dx.doi.org/10.1109/LAWP.2014.2360710]

[6]   S.F. Jilani, H. Ur-Rahman, and M.N. Iqbal, "Novel star-shaped fractal design of rectangular patch antenna for improved gain and bandwidth", *2013 IEEE Antennas and Propagation Society*

International Symposium (APSURSI), pp. 1486-1487, 2013.
[http://dx.doi.org/10.1109/APS.2013.6711402]

[7]   L.-Y. Tseng, and T.-Y. Han, "An evolutionary design method using genetic local search algorithm to obtain broad/dual-band characteristics for circular polarization slot antennas", *IEEE Trans. Antenn. Propag.*, vol. 58, no. 5, pp. 1449-1456, 2010.
[http://dx.doi.org/10.1109/TAP.2010.2044312]

[8]   K. Ito, K. Saito, and M. Takahashi, "Small antennas for medical applications", *2007 International workshop on Antenna Technology: Small and Smart Antennas Metamaterials and Applications*, pp. 116-119, 2007.

[9]   C.J. Sánchez-Fernández, O. Quevedo-Teruel, J. Requena-Carrión, L. Inclán-Sánchez, and E. Rajo-Iglesias, "Dual-band microstrip patch antenna based on short-circuited ring and spiral resonators for implantable medical devices", *IET Microw. Antennas Propag.*, vol. 4, no. 8, p. 1048, 2010.
[http://dx.doi.org/10.1049/iet-map.2009.0594]

[10]   A.Z. Hood, and E. Topsakal, "Particle swarm optimization for dual-band implantable antennas", *Proceedings of 2007 IEEE Antennas and Propagation Society International Symposium*, pp. 3209-3212, 2007.
[http://dx.doi.org/10.1109/APS.2007.4396219]

[11]   G.A. Conway, and W.G. Scanlon, "Antennas for over-body-surface communication at 2.45 GHz", *IEEE Trans. Antenn. Propag.*, vol. 57, no. 4, pp. 844-855, 2009.
[http://dx.doi.org/10.1109/TAP.2009.2014525]

[12]   P. Soontornpipit, C.M. Furse, and Y.C. Chung, "Design of implantable microstrip antenna for communication with medical implants", *IEEE Trans. Microw. Theory Tech.*, vol. 52, no. 8, pp. 1944-1951, 2004.
[http://dx.doi.org/10.1109/TMTT.2004.831976]

[13]   P. Tang, and P. Wahid, "Hexagonal fractal multiband antenna", *Proceedings of IEEE Antennas and Propagation Society International Symposium*, pp. 554-557, 2002.
[http://dx.doi.org/10.1109/APS.2002.1017045]

[14]   D.H. Wqrner, and S. Ganguly, "An overview of fractal antenna engineering research", *IEEE Antennas Propag. Mag.*, vol. 45, no. 1, pp. 38-57, 2003.
[http://dx.doi.org/10.1109/MAP.2003.1189650]

[15]   N. Song, K. Chin, D. Boon Liang, and M. Anyi, "Design of broadband dual-frequency microstrip patch antenna with modified Sierpinski fractal geometry", *Proceedings of 2006 10th IEEE Singapore International Conference on Communication Systems*, pp. 1-5, 2006.
[http://dx.doi.org/10.1109/ICCS.2006.301376]

[16]   C. Puente, J. Romeu, R. Pous, X. Garcia, and F. Benitez, "Fractal multiband antenna based on the Sierpinski gasket", *Electron. Lett.*, vol. 32, no. 1, p. 1, 1996.
[http://dx.doi.org/10.1049/el:19960033]

[17]   C. Puente-Baliarda, J. Romeu, R. Pous, and A. Cardama, "On the behavior of the Sierpinski multiband fractal antenna", *IEEE Trans. Antenn. Propag.*, vol. 46, no. 4, pp. 517-524, 1998.
[http://dx.doi.org/10.1109/8.664115]

[18]   D.H. Werner, R.L. Haupt, and P.L. Werner, "Fractal antenna engineering: the theory and design of fractal antenna arrays", *IEEE Antennas Propag. Mag.*, vol. 41, no. 5, pp. 37-58, 1999.
[http://dx.doi.org/10.1109/74.801513]

[19]   S.R. Best, "On the radiation pattern characteristics of the Sierpinski and modified Parany gasket antennas", *IEEE Antennas Wirel. Propag. Lett.*, vol. 1, pp. 39-42, 2002.
[http://dx.doi.org/10.1109/LAWP.2002.802585]

[20]   J.P. Gianvittorio, and Y. Rahmat-Samii, "Fractal antennas: a novel antenna miniaturization technique, and applications", *IEEE Antennas Propag. Mag.*, vol. 44, no. 1, pp. 20-36, 2002.

[http://dx.doi.org/10.1109/74.997888]

[21]   S.R. Best, "A comparison of the resonant properties of small space-filling fractal antennas", *IEEE Antennas Wirel. Propag. Lett.,* vol. 2, no. 13, pp. 197-200, 2003.
[http://dx.doi.org/10.1109/LAWP.2003.819680]

[22]   P. Dehkhoda, and A. Tavakoli, "A crown square microstrip fractal antenna", *Proceedings of IEEE Antennas and Propagation Society Symposium,* pp. 2396-2399, 2004.
[http://dx.doi.org/10.1109/APS.2004.1331855]

[23]   J. Anguera, E. Martinez, C. Puente, C. Borja, and J. Soler, "Broad-band dual-frequency microstrip patch antenna with modified Sierpinski fractal geometry", *IEEE Trans. Antenn. Propag.,* vol. 52, no. 1, pp. 66-73, 2004.
[http://dx.doi.org/10.1109/TAP.2003.822433]

[24]   M.K.A. Rahim, M.Z.A.A. Aziz, and N. Abdullah, "Microstrip Sierpinski carpet antenna using transmission line feeding", *Proceedings of 2005 Asia-Pacific Microwave Conference,* pp. 1-4, 2005.
[http://dx.doi.org/10.1109/APMC.2005.1606382]

[25]   M. Ding, R. Jin, J. Geng, Q. Wu, and W. Wang, "Design of a CPW-fed ultra wideband crown circular fractal antenna", *Proceedings of 2006 IEEE Antennas and Propagation Society International Symposium,* pp. 2049-2052, 2006.
[http://dx.doi.org/10.1109/APS.2006.1710983]

[26]   J. Huang, F. Shan, and J. She, "A novel multiband and broadband fractal patch antenna", *Prog. Electromagn. Res. Symp.,* vol. 2, no. 1, pp. 57-59, 2006.
[http://dx.doi.org/10.2529/PIERS051007103829]

[27]   J. Anguera, E. Martinez-Ortigosa, C. Puente, C. Borja, and J. Soler, "Broadband triple-frequency microstrip patch radiator combining a dual-band modified Sierpinski fractal and a monoband antenna", *IEEE Trans. Antenn. Propag.,* vol. 54, no. 11, pp. 3367-3373, 2006.
[http://dx.doi.org/10.1109/TAP.2006.884209]

[28]   X. Guangguo, and Z. Shouzheng, "Novel fractal and MEMS fractal antennas", *Proceedings of 2007 International Conference on Microwave and Millimeter Wave Technology,* pp. 1-4, 2007.

[29]   D.D. Krishna, A.R. Chandran, and C.K. Aanandan, "A compact dual frequency antenna with Sierpinski gasket based slots", *Proceedings of 2007 European Microwave Conference,* pp. 1078-1080, 2007.
[http://dx.doi.org/10.1109/EUMC.2007.4405384]

[30]   K.C. Hwang, "A modified Sierpinski fractal antenna for multiband application", *IEEE Antennas Wirel. Propag. Lett.,* vol. 6, pp. 357-360, 2007.
[http://dx.doi.org/10.1109/LAWP.2007.902045]

[31]   R. Kumar, P. Malathi, and J.P. Shinde, "Design of miniaturized fractal antenna", *Proceedings of 2007 European Microwave Conference,* pp. 474-477, 2007.
[http://dx.doi.org/10.1109/EUMC.2007.4405230]

[32]   A. Azari, and J. Rowhani, "Ultra wideband fractal microstrip antenna design", *Prog. Electromagn. Res. C. Pier C,* vol. 2, pp. 7-12, 2008.
[http://dx.doi.org/10.2528/PIERC08031005]

[33]   M. Naghshvarian-Jahromi, "Novel wideband planar fractal monopole antenna", *IEEE Trans. Antenn. Propag.,* vol. 56, no. 12, pp. 3844-3849, 2008.
[http://dx.doi.org/10.1109/TAP.2008.2007393]

[34]   R. Mishra, R. Ghatak, and D. Poddar, "Design formula for Sierpinski gasket pre-fractal planar-monopole antennas", *IEEE Antennas Propag. Mag.,* vol. 50, no. 3, pp. 104-107, 2008.
[http://dx.doi.org/10.1109/MAP.2008.4563575]

[35]   W. Yong, and L. Shaobin, "A new modified crown square fractal antenna", *Proceedings of 2008 International Conference on Microwave and Millimeter Wave Technology,* pp. 400-402, 2008.

[http://dx.doi.org/10.1109/ICMMT.2008.4540400]

[36]   B. Mirzapour, and H.R. Hassani, "Size reduction and bandwidth enhancement of snowflake fractal antenna", *IET Microw. Antennas Propag.,* vol. 2, no. 2, pp. 180-187, 2008.
[http://dx.doi.org/10.1049/iet-map:20070133]

[37]   W.J. Krzysztofik, "Modified Sierpinski fractal monopole for ISM-bands handset applications", *IEEE Trans. Antenn. Propag.,* vol. 57, no. 3, pp. 606-615, 2009.
[http://dx.doi.org/10.1109/TAP.2009.2013416]

[38]   B. Manimegalai, S. Raju, and V. Abhaikumar, "A multifractal cantor antenna for multiband wireless applications", *IEEE Antennas Wirel. Propag. Lett.,* vol. 8, pp. 359-362, 2009.
[http://dx.doi.org/10.1109/LAWP.2008.2000828]

[39]   L. Cao, S. Yan, and H. Yang, "Study and design of a modified fractal antenna for RFID applications", *Proceedings of 2009 ISECS International Colloquium on Computing, Communication, Control, and Management,* pp. 8-11, 2009.
[http://dx.doi.org/10.1109/CCCM.2009.5268163]

[40]   J. Gemio, J. Parron, and J. Soler, "Multiband antenna for WLAN applications using a fractal-based ground plane", *Proceedings of European Conference on Antennas and Propagation,* pp. 362-365, 2009.

[41]   A. Aggarwal, and M.V. Kartikeyan, "Pythagoras Tree: a fractal patch antenna for multi-frequency and ultra-wide bandwidth operations", *Prog. Electromagn. Res. C. Pier C,* vol. 16, pp. 25-35, 2010.
[http://dx.doi.org/10.2528/PIERC10062206]

[42]   S.A. Kumar, and T.K. Sreeja, "A modified fractal antenna for multiband applications", *Proceedings of IEEE International Conference on Communication Control and Computing Technologies,* pp. 47-51, 2010.Ramanathapuram

[43]   S.R. Anoop, K.K. Ajayan, M.R. Baiju, and V. Krishnakumar, "Multiband behavioural analysis of a higher order fractal patch antenna", *Proceedings of International Congress on Ultra Modern Telecommunications and Control Systems,* pp. 823-827, 2010.
[http://dx.doi.org/10.1109/ICUMT.2010.5676540]

[44]   F.J. Jibrael, and M.H. Hammed, "A new multiband patch microstrip plusses fractal antenna for wireless applications", *J. Eng. Appl. Sci. (Asian Res. Publ. Netw.),* vol. 5, no. 8, pp. 17-21, 2010.

[45]   A. Azari, "A new super wideband fractal microstrip antenna", *IEEE Trans. Antenn. Propag.,* vol. 59, no. 5, pp. 1724-1727, 2011.
[http://dx.doi.org/10.1109/TAP.2011.2128294]

[46]   N. Bayatmaku, P. Lotfi, M. Azarmanesh, and S. Soltani, "Design of simple multiband patch antenna for mobile communication applications using new E-shape fractal", *IEEE Antennas Wirel. Propag. Lett.,* vol. 10, pp. 873-875, 2011.
[http://dx.doi.org/10.1109/LAWP.2011.2165195]

[47]   Y.K. Choukiker, S. Rai, and S.K. Behera, "Modified half-circle fractal antenna using DC theorem for 2.4/5.2 GHz WLAN application", *Proceedings of 2011 National Conference on Communications (NCC),* pp. 1-4, 2011.
[http://dx.doi.org/10.1109/NCC.2011.5734744]

[48]   P.J. Soh, G.A.E. Vandenbosch, S.L. Ooi, and M.R.N. Husna, "Wearable dual-band Sierpinski fractal PIFA using conductive fabric", *Electron. Lett.,* vol. 47, no. 6, p. 365, 2011.
[http://dx.doi.org/10.1049/el.2010.3525]

[49]   Y.J. Sung, "Bandwidth enhancement of a wide slot using fractal-shaped Sierpinski", *IEEE Trans. Antenn. Propag.,* vol. 59, no. 8, pp. 3076-3079, 2011.
[http://dx.doi.org/10.1109/TAP.2011.2158942]

[50]   E.C. Lee, P.J. Soh, N.B.M. Hashim, G.A.E. Vandenbosch, V. Volski, I. Adam, H. Mirza, and M.Z.A.A. Aziz, "Design and fabrication of a flexible Minkowski fractal antenna for VHF

applications", *Proceedings of European Conference on Antennas and Propagation,* pp. 521-524, 2011.Rome

[51]   J. Pourahmadazar, C. Ghobadi, and J. Nourinia, "Novel modified Pythagorean tree fractal monopole antennas for UWB applications", *IEEE Antennas Wirel. Propag. Lett.,* vol. 10, pp. 484-487, 2011.
[http://dx.doi.org/10.1109/LAWP.2011.2154354]

[52]   M. Naghshvarian Jahromi, A. Falahati, and R.M. Edwards, "Bandwidth and impedance-matching enhancement of fractal monopole antennas using compact grounded coplanar waveguide", *IEEE Trans. Antenn. Propag.,* vol. 59, no. 7, pp. 2480-2487, 2011.
[http://dx.doi.org/10.1109/TAP.2011.2152321]

[53]   R. Kumar, and P.B. Nikam, "A modified ground apollonian ultra wideband fractal antenna and its backscattering", *AEU Int. J. Electron. Commun.,* vol. 66, no. 8, pp. 647-654, 2012.
[http://dx.doi.org/10.1016/j.aeue.2011.12.002]

[54]   H. Oraizi, and S. Hedayati, "Circularly polarized multiband microstrip antenna using the square and Giuseppe peano fractals", *IEEE Trans. Antenn. Propag.,* vol. 60, no. 7, pp. 3466-3470, 2012.
[http://dx.doi.org/10.1109/TAP.2012.2196912]

[55]   R. Kumar, D. Magar, and K. Kailas Sawant, "On the design of inscribed triangle circular fractal antenna for UWB applications", *AEU Int. J. Electron. Commun.,* vol. 66, no. 1, pp. 68-75, 2012.
[http://dx.doi.org/10.1016/j.aeue.2011.05.003]

[56]   H. Oraizi, and S. Hedayati, "Miniaturization of microstrip antennas by the novel application of the Giuseppe peano fractal geometries", *IEEE Trans. Antenn. Propag.,* vol. 60, no. 8, pp. 3559-3567, 2012.
[http://dx.doi.org/10.1109/TAP.2012.2201070]

[57]   O.O. Olaode, W.D. Palmer, and W.T. Joines, "Effects of meandering on dipole antenna resonant frequency", *IEEE Antennas Wirel. Propag. Lett.,* vol. 11, pp. 122-125, 2012.
[http://dx.doi.org/10.1109/LAWP.2012.2184255]

[58]   D. Li, and J. Mao, "A Koch-like sided fractal bow-tie dipole antenna", *IEEE Trans. Antenn. Propag.,* vol. 60, no. 5, pp. 2242-2251, 2012.
[http://dx.doi.org/10.1109/TAP.2012.2189719]

[59]   R. Ghatak, A. Karmakar, and D.R. Poddar, "Hexagonal boundary Sierpinski carpet fractal shaped compact ultrawideband antenna with band rejection functionality", *AEU Int. J. Electron. Commun.,* vol. 67, no. 3, pp. 250-255, 2013.
[http://dx.doi.org/10.1016/j.aeue.2012.08.007]

[60]   Y-Y. Xu, Y. Xu, J. Hu, and W-Y. Yin, "Design of a novel reconfigurable Sierpinski fractal graphene antenna operating at THz band", *Proceedings of 2013 IEEE Antennas and Propagation Society International Symposium (APSURSI),* pp. 574-575, 2013.
[http://dx.doi.org/10.1109/APS.2013.6710947]

[61]   M.A. Dorostkar, M.T. Islam, and R. Azim, "Design of a novel super wide band circular-hexagonal fractal antenna", *Electromagn. waves,* vol. 139, pp. 229-245, 2013.
[http://dx.doi.org/10.2528/PIER13030505]

[62]   R. Kumar, and P.N. Choubey, "Design of pentagonal circular fractal antenna with and without notched-band characteristics", *Microw. Opt. Technol. Lett.,* vol. 55, no. 2, pp. 430-434, 2013.
[http://dx.doi.org/10.1002/mop.27316]

[63]   M. Naser-Moghadasi, R.A. Sadeghzadeh, T. Sedghi, T. Aribi, and B.S. Virdee, "UWB CPW-fed fractal patch antenna with band-notched function employing folded T-shaped element", *IEEE Antennas Wirel. Propag. Lett.,* vol. 12, pp. 504-507, 2013.
[http://dx.doi.org/10.1109/LAWP.2013.2256455]

[64]   C. Varadhan, J.K. Pakkathillam, M. Kanagasabai, R. Sivasamy, R. Natarajan, and S.K. Palaniswamy, "Triband antenna structures for RFID systems deploying fractal geometry", *IEEE Antennas Wirel.*

Propag. Lett., vol. 12, pp. 437-440, 2013.
[http://dx.doi.org/10.1109/LAWP.2013.2254458]

[65]   A. Karmakar, U. Banerjee, R. Ghatak, and D.R. Poddar, "Design and analysis of fractal based UWB monopole antenna", *Proceedings of 2013 National Conference on Communications (NCC),* pp. 1-5, 2013.
[http://dx.doi.org/10.1109/NCC.2013.6487975]

[66]   S. Shrestha, S-J. Han, S-K. Noh, S. Kim, H-B. Kim, and D-Y. Choi, "Design of modified Sierpinski fractal based miniaturized patch antenna", *Proceedings of the International Conference on Information Networking 2013 (ICOIN),* pp. 274-279, 2013.
[http://dx.doi.org/10.1109/ICOIN.2013.6496389]

[67]   T.N. Thi, K.C. Hwang, and H.B. Kim, "Dual-band circularly-polarised Spidron fractal microstrip patch antenna for Ku-band satellite communication applications", *Electron. Lett.,* vol. 49, no. 7, pp. 444-445, 2013.
[http://dx.doi.org/10.1049/el.2012.2973]

[68]   H. Fallahi, and Z. Atlasbaf, "Study of a class of UWB CPW-fed monopole antenna with fractal elements", *IEEE Antennas Wirel. Propag. Lett.,* vol. 12, pp. 1484-1487, 2013.
[http://dx.doi.org/10.1109/LAWP.2013.2289868]

[69]   J. Kizhekke Pakkathillam, M. Kanagasabai, C. Varadhan, and P. Sakthivel, "A novel fractal antenna for UHF near-field RFID readers", *IEEE Antennas Wirel. Propag. Lett.,* vol. 12, pp. 1141-1144, 2013.
[http://dx.doi.org/10.1109/LAWP.2013.2281069]

[70]   G. Liu, L. Xu, and Z. Wu, "Dual-band microstrip RFID antenna with tree-like fractal structure", *IEEE Antennas Wirel. Propag. Lett.,* vol. 12, pp. 976-978, 2013.
[http://dx.doi.org/10.1109/LAWP.2013.2276933]

[71]   G. Liu, L. Xu, and Z. Wu, "Miniaturised wideband circularly-polarised log-periodic Koch fractal antenna", *Electron. Lett.,* vol. 49, no. 21, pp. 1315-1316, 2013.
[http://dx.doi.org/10.1049/el.2013.2418]

[72]   Y.K. Choukiker, and S.K. Behera, "Compact sectoral fractal planar monopole antenna for wideband wireless applications", *Microw. Opt. Technol. Lett.,* vol. 56, no. 5, pp. 1073-1076, 2014.
[http://dx.doi.org/10.1002/mop.28272]

[73]   H.L. Dholakiya, and D.A. Pujara, "Improving the bandwidth of a microstrip antenna with a circular-shaped fractal slot", *Microw. Opt. Technol. Lett.,* vol. 55, no. 4, pp. 786-789, 2013.
[http://dx.doi.org/10.1002/mop.27424]

[74]   M.E.B. Jalil, M.K. Abd Rahim, N.A. Samsuri, N.A. Murad, H.A. Majid, K. Kamardin, and M. Azfar Abdullah, "Fractal Koch multiband textile antenna performance with bending, wet conditions and on the human body", *Electromagn. waves,* vol. 140, pp. 633-652, 2013.
[http://dx.doi.org/10.2528/PIER13041212]

[75]   H.X. Xu, G.M. Wang, J.G. Liang, M.Q. Qi, and X. Gao, "Compact circularly polarized antennas combining meta-surfaces and strong space-filling meta-resonators", *IEEE Trans. Antenn. Propag.,* vol. 61, no. 7, pp. 3442-3450, 2013.
[http://dx.doi.org/10.1109/TAP.2013.2255855]

[76]   G. Wang, D. Shen, and X. Zhang, "An UWB antenna using modified Sierpinski-carpet Fractal Antenna", *Proceedings of 2013 IEEE Antennas and Propagation Society International Symposium (APSURSI),* pp. 216-217, 2013.
[http://dx.doi.org/10.1109/APS.2013.6710769]

[77]   F.B. Zarrabi, A.M. Shire, M. Rahimi, and N.P. Gandji, "Ultra-wideband tapered patch antenna with fractal slots for dual notch application", *Microw. Opt. Technol. Lett.,* vol. 56, no. 6, pp. 1344-1348, 2014.
[http://dx.doi.org/10.1002/mop.28332]

[78] M. Jalali, and T. Sedghi, "Very compact UWB CPW-fed fractal antenna using modified ground plane and unit cells", *Microw. Opt. Technol. Lett.,* vol. 56, no. 4, pp. 851-854, 2014.
[http://dx.doi.org/10.1002/mop.28194]

[79] D.Y. Choi, S. Shrestha, J.J. Park, and S.K. Noh, "Design and performance of an efficient rectenna incorporating a fractal structure", *Int. J. Commun. Syst.,* vol. 27, no. 4, pp. 661-679, 2014.
[http://dx.doi.org/10.1002/dac.2587]

[80] V.V. Reddy, and N.V.S.N. Sarma, "Compact circularly polarized asymmetrical fractal boundary microstrip antenna for wireless applications", *IEEE Antennas Wirel. Propag. Lett.,* vol. 13, pp. 118-121, 2014.
[http://dx.doi.org/10.1109/LAWP.2013.2296951]

[81] V.V. Reddy, and N.V.S.N. Sarma, "Triband circularly polarized Koch fractal boundary microstrip antenna", *IEEE Antennas Wirel. Propag. Lett.,* vol. 13, pp. 1057-1060, 2014.
[http://dx.doi.org/10.1109/LAWP.2014.2327566]

[82] A.S. Abd El-Hameed, D.A. Salem, E.A. Abdallah, and E.A. Hashish, "Crossbar fractal quasi self-complementary UWB antenna", *Proceedings of 2014 IEEE Antennas and Propagation Society International Symposium (APSURSI),* pp. 219-220, 2014.
[http://dx.doi.org/10.1109/APS.2014.6904441]

[83] S. Subramaniam, and S. Dhar, "K., Patra, B. Gupta, L. Osman, K. Zeouga, and A. Gharsallah, "Miniaturization of wearable electro-textile antennas using Minkowski fractal geometry", *Proceedings of 2014 IEEE Antennas and Propagation Society International Symposium (APSURSI),* pp. 309-310, 2014.
[http://dx.doi.org/10.1109/APS.2014.6904486]

[84] L. Li, Z. Wu, K. Li, S. Yu, X. Wang, T. Li, G. Li, X. Chen, and H. Zhai, "Frequency-reconfigurable quasi-Sierpinski antenna integrating with dual-band high-impedance surface", *IEEE Trans. Antenn. Propag.,* vol. 62, no. 9, pp. 4459-4467, 2014.
[http://dx.doi.org/10.1109/TAP.2014.2331992]

[85] C. Raviteja, C. Varadhan, M. Kanagasabai, A.K. Sarma, and S. Velan, "A fractal-based circularly polarized UHF RFID reader antenna", *IEEE Antennas Wirel. Propag. Lett.,* vol. 13, pp. 499-502, 2014.
[http://dx.doi.org/10.1109/LAWP.2014.2308953]

[86] M.M.A. Kumar, A. Patnaik, and C.G. Christodoulou, "Design and testing of a multifrequency antenna with a reconfigurable feed", *IEEE Antennas Wirel. Propag. Lett.,* vol. 13, pp. 730-733, 2014.
[http://dx.doi.org/10.1109/LAWP.2014.2315433]

[87] W.C. Weng, and C.L. Hung, "An H-fractal antenna for multiband applications", *IEEE Antennas Wirel. Propag. Lett.,* vol. 13, pp. 1705-1708, 2014.
[http://dx.doi.org/10.1109/LAWP.2014.2351618]

[88] S. Chatterjee, A. Majumder, R. Ghatak, and D.R. Poddar, "Wide impedance and pattern bandwidth realization using fractal slotted array antenna", *IEEE Trans. Antenn. Propag.,* vol. 62, no. 8, pp. 4049-4056, 2014.
[http://dx.doi.org/10.1109/TAP.2014.2322887]

[89] C. Zhou, J. Fu, S. Lin, and Q. Wu, "Broadband circularly polarized antenna with a tree fractal wide-slot and a L-shaped strip", *Proceedings of 2014 IEEE Antennas and Propagation Society International Symposium (APSURSI),* pp. 221-222, 2014.
[http://dx.doi.org/10.1109/APS.2014.6904442]

[90] S. Costanzo, and F. Venneri, "Miniaturized fractal reflectarray element using fixed-size patch", *IEEE Antennas Wirel. Propag. Lett.,* vol. 13, pp. 1437-1440, 2014.
[http://dx.doi.org/10.1109/LAWP.2014.2341032]

[91] S. Tripathi, A. Mohan, and S. Yadav, "A compact Koch fractal UWB MIMO antenna with WLAN

band-rejection", *IEEE Antennas Wirel. Propag. Lett.,* vol. 14, pp. 1565-1568, 2015.
[http://dx.doi.org/10.1109/LAWP.2015.2412659]

[92]   S. Tripathi, A. Mohan, and S. Yadav, "A compact octagonal-shaped fractal UWB antenna with
       Sierpinski fractal geometry", *Microw. Opt. Technol. Lett.,* vol. 57, no. 3, pp. 570-574, 2015.
       [http://dx.doi.org/10.1002/mop.28901]

[93]   S. Singhal, T. Goel, and A. Kumar Singh, "Inner tapered tree-shaped fractal antenna for UWB
       applications", *Microw. Opt. Technol. Lett.,* vol. 57, no. 3, pp. 559-567, 2015.
       [http://dx.doi.org/10.1002/mop.28900]

[94]   J. Kizhekke Pakkathillam, and M. Kanagasabai, "Circularly polarized broadband antenna deploying
       fractal slot geometry", *IEEE Antennas Wirel. Propag. Lett.,* vol. 14, pp. 1286-1289, 2015.
       [http://dx.doi.org/10.1109/LAWP.2015.2402286]

[95]   A. Altaf, Y. Yang, K.Y. Lee, and K.C. Hwang, "Circularly polarized spidron fractal dielectric
       resonator antenna", *IEEE Antennas Wirel. Propag. Lett.,* vol. 14, pp. 1806-1809, 2015.
       [http://dx.doi.org/10.1109/LAWP.2015.2427373]

[96]   T. Cai, G.M. Wang, X.F. Zhang, and J.P. Shi, "Low-profile compact circularly-polarized antenna
       based on fractal metasurface and fractal resonator", *IEEE Antennas Wirel. Propag. Lett.,* vol. 14, pp.
       1072-1076, 2015.
       [http://dx.doi.org/10.1109/LAWP.2015.2394452]

[97]   A. Amini, H. Oraizi, and M.A. Chaychi zadeh, "Miniaturized UWB log-periodic square fractal
       antenna", *IEEE Antennas Wirel. Propag. Lett.,* vol. 14, pp. 1322-1325, 2015.
       [http://dx.doi.org/10.1109/LAWP.2015.2411712]

[98]   G. Geetha, S.K. Palaniswamy, M. Alsath, M. Kanagasabai, and T.R. Rao, "Compact and flexible
       monopole antenna for ultra-wideband applications deploying fractal geometry", *J. Electr. Eng.
       Technol.,* vol. 13, no. 1, pp. 400-405, 2018.

[99]   F. Mokhtari-Koushyar, P.M. Grubb, M.Y. Chen, and R.T. Chen, "A miniaturized tree-shaped fractal
       antenna printed on a flexible substrate", *IEEE Antennas Propag. Mag.,* vol. 61, no. 3, pp. 60-66, 2019.
       [http://dx.doi.org/10.1109/MAP.2019.2907878]

[100] A. Al-Sehemi, A. Al-Ghamdi, N. Dishovsky, G. Atanasova, and N. Atanasov, "Flexible
       polymer/fabric fractal monopole antenna for wideband applications", *IET Microw. Antennas Propag.,*
       vol. 15, no. 1, pp. 80-92, 2021.
       [http://dx.doi.org/10.1049/mia2.12016]

[101] S.G. Kirtania, A.W. Elger, M.R. Hasan, A. Wisniewska, K. Sekhar, T. Karacolak, and P.K. Sekhar,
       "Flexible antennas: a review", *Micromachines (Basel),* vol. 11, no. 9, p. 847, 2020.
       [http://dx.doi.org/10.3390/mi11090847] [PMID: 32933077]

[102] P. Kumar, T. Ali, and A. Sharma, "Flexible substrate based printed wearable antennas for wireless
       body area networks medical applications (review)", *Radioelectron. Commun. Syst.,* vol. 64, no. 7, pp.
       337-350, 2021.
       [http://dx.doi.org/10.3103/S0735272721070013]

[103] V. Jain, and B.S. Dhaliwal, "Investigations on the design of compact flexible wearable fractal patch
       antenna for body area networks applications", *Wirel. Pers. Commun.,* vol. 126, no. 2, pp. 1443-1458,
       2022.
       [http://dx.doi.org/10.1007/s11277-022-09800-0]

Bio-inspired Computing Techniques and their Applications in Antennas

Abstract: This chapter is dedicated to bio-inspired computing techniques and their applications in antennas. The working principles of ANN, ANN Ensemble, GA, PSO, and BFO are described, and some hybrid bio-inspired computing techniques are also discussed. The literature survey related to the applications of bio-inspired computing optimization techniques in antennas is given in this chapter. The limitations of the existing bio-inspired computing techniques are highlighted. The existing applications of bio-inspired computing techniques in fractal antennas are also reviewed in this chapter.

Keywords: Bacterial foraging optimization, Fractal antenna, Genetic algorithms, Particle swarm optimization, Sierpinski gasket.

INTRODUCTION

Many bio-inspired computing techniques have been proposed by researchers in the past two decades that imitate genetic evolution, the human brain or the behavior of biological individuals in the natural world. Bio-inspired computing can tolerate uncertainty, vagueness, partial truth, and approximation. The bio-inspired computing techniques make use of these tolerances to achieve tractability, robustness and low solution cost [1]. A precise analytical model is required in traditional computing; the computation time is generally very large. Also, in most of the cases, the analytical models are applicable to ideal situations, and real-world problems exist in non-ideal conditions. Bio-inspired computing techniques are a group of methods spanning several areas that come under different categories in artificial intelligence. Despite the fact that some of these algorithms are new, they have been used effectively in many optimization problems with several constraints [1]. The main constituents of bio-inspired computing techniques are ANN, GA, PSO, BFO, and SIMBO. At present, these techniques are fundamental to many antenna design solutions and have shown great promise in tackling the growing requirements of antenna engineering for improved performances, reduced size, and overall cost [2, 3]. The bio-inspired computing techniques have resulted in many useful and non-intuitive antenna design solutions and, in most of the cases, these techniques have outperformed the traditional methods. Another important reason for the suitability of bio-inspired computing techniques is that antenna synthesis and optimization problems frequently involve many parameters and a very large number of possibilities, thus making exhaustive searches impractical. The use of these techniques for fractal antennas' design is very suitable because of the non-availability of exact mathematical expressions for these antennas [3, 4].

BIO-INSPIRED COMPUTING TECHNIQUES

This section describes the working of the important constituents of the soft computing algorithm.

Artificial Neural Network (ANN)

ANNs have been employed for analysis and prediction in almost all disciplines of engineering as these can model non-linear systems very efficiently. The use of ANN models for the analysis and design of microstrip antennas is widely accepted. This is obvious from the growing count of research/academic journals' publications in this field. ANNs have been used for the analysis of antennas, synthesis of antennas, parameter estimation, *etc.*, as discussed in the following sections.

The ANNs are computational models inspired by biological neural networks. The ANNs can process information like the human brain to attain, gather and exploit experimental knowledge. ANNs are used to find patterns in data or to model complex relationships between inputs and outputs [1]. The basic computational unit of the ANN is the neuron. The ANNs are massive parallel structures of neurons arranged in the form of layers having interconnections between them. The neurons are of three types: (i) input neurons which receive inputs from the outside world, (ii) hidden neurons which receive inputs from other neurons and whose outputs are inputs for other neurons in the network, and (iii) output neurons whose output is supplied to the outside world [2]. The connecting branches of the neurons have the associated connection strength parameters which are known as the weights. The design of the ANN involves the training of the network for the specific application. During training, an input-output data set is used to adjust the weights so that the ANN produces the correct output for the input applied [4]. Once the network is successfully trained, it can be used for predicting outputs for unknown inputs.

There exist different types of ANNs, which are constructed by using different kinds of interconnections and various types of neurons. The most common type of ANN is feed-forward neural networks in which the information flows in the forward direction only, *i.e.*, there is no feedback connection. The three important forms of feed-forward ANNs are Multi-Layer Perceptron Neural Networks (MLPNN), Radial Basis Function Neural Networks (RBFNN), and General Regression Neural Networks (GRNN) [1]. These three forms are very popular in the field of antennas.

Multi-Layer Perceptron Neural Networks (MLPNN)

MLPNN is a widely used neural network structure in antenna applications. It consists of multiple layers of neurons that are connected in a feed-forward manner [4]. The base structure of an MLPNN is shown in Fig. (**2.1**), which depicts that the network has K layers. The first layer is the input layer, and the last layer (K^{th} layer) is the output layer. The other layers, *i.e.*, layers from *2 to K-1*, are hidden layers. Each layer has a number of neurons [5]. In the input and output layers, the numbers of neurons are equal to the number of inputs and outputs, respectively. The neurons in the hidden layers are selected using a trial-and-error process, so that accurate output is obtained. Each neuron processes the inputs to produce output using a function called the activation function.

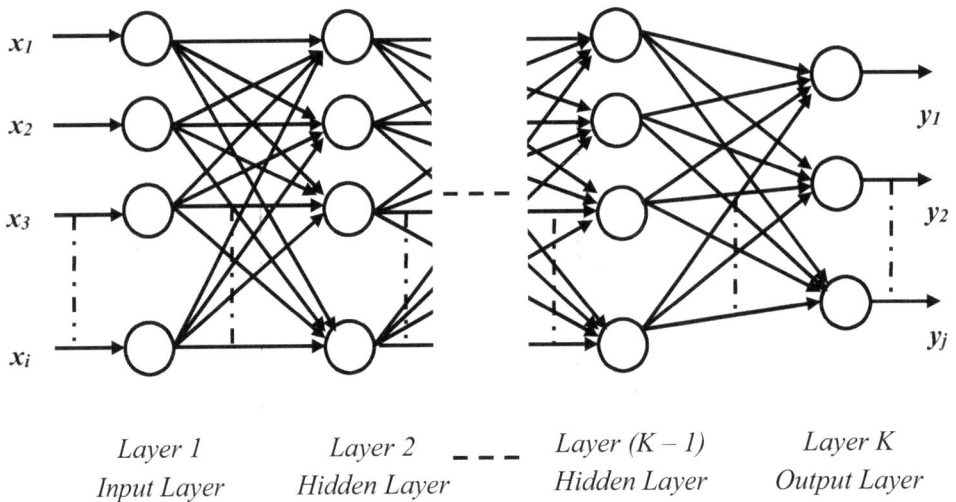

| Layer 1 | Layer 2 | --- | Layer (K – 1) | Layer K |
| Input Layer | Hidden Layer | | Hidden Layer | Output Layer |

Fig. (2.1). Basic Structure of MLPNN [6].

The input neurons normally use a relay activation function and simply relay (pass) the external inputs to the hidden layer neurons. The most commonly used activation function for hidden layer neurons is the sigmoid function defined as in the equation given below [2]:

$$\sigma(r) = \frac{1}{1+e^{-r}} \qquad\qquad (2.1)$$

The connection between the units in subsequent layers has an associated weight which is computed during training using the error backpropagation algorithm [2]. The back-propagation learning algorithm consists of two steps. In the first step, the input signal is passed forward through each layer of the network. The actual output of the network is calculated, and this output signal is subtracted from desired output signal to generate an error signal. In the second step, the error is fed backward through the network from the output layer through the hidden layers. The synaptic weights between each layer are updated based on the computational error. These two steps are repeated for each input-output pair of training data, and the process is iterated a number of times until the output value converges to a desired solution [2].

Radial Basis Function Neural Networks (RBFNN)

RBFNN has several special characteristics, like simple architecture and faster performance [7]. It is a special neural network with a single hidden layer. The hidden layer neurons use Radial Basis Function (RBF) to apply a nonlinear transformation from the input space to the hidden space [8]. The nonlinear basis function is a function of the normalized radial distance between the input vector and the weight vector. The most widely used form of RBF is the Gaussian function, and it is given in the equation below [8]:

$$\Phi_j(x) = \exp\left(\frac{-\|x - \mu_j\|^2}{\sigma_j^2}\right) \qquad\qquad (2.2)$$

where x is a d-dimensional input vector with elements x_i, and μ_j is the vector determining the centre of basis function Φ_j and has elements μ_{ji}. The symbol σ_j denotes the width parameter whose value is determined during training. The RBFNN involves a two-stage training procedure. In the first stage, the parameters governing the basis functions, *i.e.*, centers (μ_j) and widths (σ_j), are determined using relatively fast, unsupervised methods. The unsupervised training methods use only the input data and not the target data. In the second stage of training, the weights of the output layer are determined [8]. For these networks, the output layer is a linear layer, so fast linear supervised methods are used. Due to this reason, these networks have faster performances [7]. The number of neurons in all layers of RBFNN, *i.e.*, input, output, and hidden layers, is determined by the dimensions and number of input-output data sets. The basic structure of RBFNN is shown in Fig. (**2.2**).

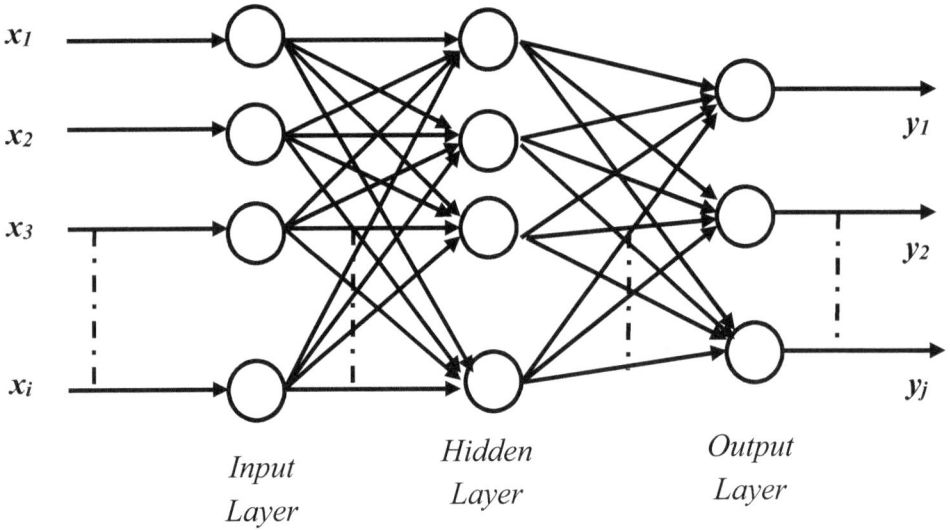

Fig. (2.2). Basic Structure of RBFNN [6].

General Regression Neural Networks (GRNN)

GRNN is a one-pass learning algorithm network with a highly parallel structure [9]. It is a memory-based network that provides an estimate of continuous variables and converges to an underlying linear or nonlinear regression surface. Nonlinear regression analysis forms the theoretical base of GRNN. If the joint probability density function of random variable x and y is $f(x, y)$ and the measured value of x is X, then the regression of y with respect to X (which is also called conditional mean) is as follows [10]:

$$\hat{Y}(X) = E\left[\frac{y}{X}\right] = \frac{\int_{-\infty}^{\infty} y.f(X,y)dy}{\int_{-\infty}^{\infty} f(X,y)dy} \tag{2.3}$$

where $\hat{Y}(X)$ is the prediction output of Y when the input is X.

For GRNN, [11] has shown that $\hat{Y}(X)$ can be calculated by using equations (2.4) and (2.5):

$$\hat{Y}(X) = \frac{\sum_{i=1}^{N} Y^i \exp(^{-D_i^2}/_{2\sigma^2})}{\sum_{i=1}^{N} \exp(^{-D_i^2}/_{2\sigma^2})} \tag{2.4}$$

$$D_i^2 = (X - X^i)^T (X - X^i) \qquad (2.5)$$

where X^i and Y^i are sample values of random variables x and y, N is number of samples in training set, σ is smoothing parameter and D_i^2 is Euclidean matrix.

The estimate $\widehat{Y}(X)$ can be visualized as a weighted average of all of the observed values, Y^i, where each observed value is weighted exponentially according to its Euclidean distance from X.

The basic structure of GRNN is shown in Fig. (**2.3**). The input units are merely distribution units, which provide all inputs (scaled measurement variables) to all the neurons on the second layer called pattern units. The number of neurons in the input layer and output layer of GRNN is equal to the dimensions of the input data and output data, respectively. The pattern unit consists of neurons equal to a number of samples in the training set, and each pattern unit is dedicated to one cluster centre. When a new vector, X, is entered into the network, it is subtracted from the stored vector representing each cluster center. The sum of squares or the absolute values of the differences is calculated and fed into a non-linear activation function which is generally an exponential function.

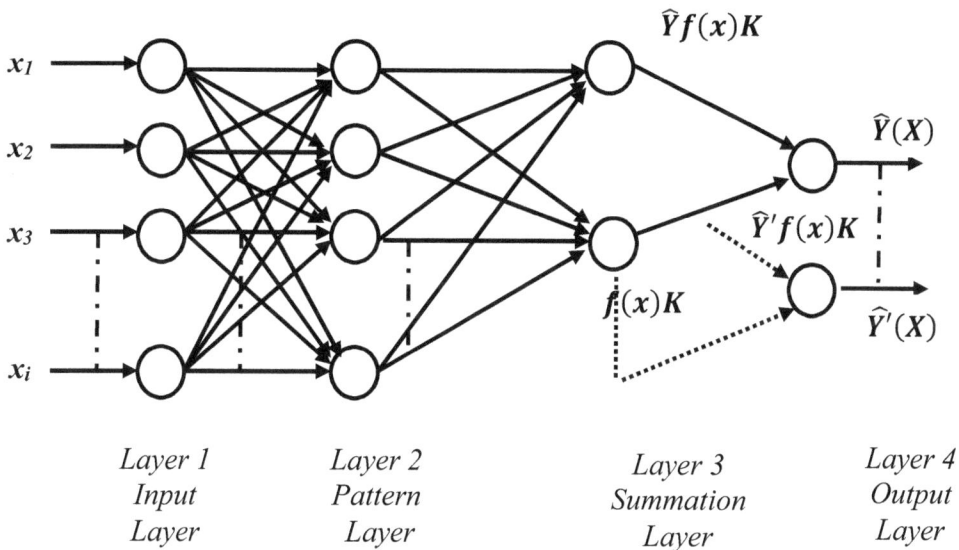

Fig. (2.3). Basic Structure of GRNN [6].

The outputs of pattern units are passed on to third-layer neurons known as summation units. There are two types of neurons in summations units: numerator neurons whose number is equal to output dimensions and one denominator neuron. The first type generates an estimate of $f(X)K$ by adding the outputs of pattern units weighted by the number of observations each cluster centre represents. The second type generates an estimate of $f(X)K$ and multiplies each value from a pattern unit by the sum of samples associated with the cluster center X^i. K is a constant determined by the Parzen window. The desired estimate of Y is yielded by the output unit by dividing $\hat{Y}f(x)K$ by $\hat{Y}f(x)K$ [11].

ANN Ensemble

In the standard ANN design approach, the final ANN architecture, its weight matrix and other tuning parameters are decided by performing trial & error iterations to find the best network, and then all future outputs of this best network are trusted [2]. In ANN ensemble design, the concept is to retain the complete set of networks (or at least a selected subset) and then employ all of them to generate the outputs which are combined through some methods [12]. This set (or the subset as the case may be) of ANNs employed in parallel for the same input-output function with outputs combined is known as an ANN ensemble [13]. The selection of weights of ANN is an optimization problem with numerous local minima. During the training of ANN, the weights are initialized from different random points; hence, various trial runs will result in optimal weights which are different from each other. These networks having different values of the weights evolved during trial runs will be making errors on different parts of the input space. So, the collective output of the ensemble is less likely to be in error than the output of any of the individual networks [14]. The several other reasons [15, 16] to support the superiority of ANN ensembles over individual networks are as follows:

- Response of individual networks is not unique, *i.e.*, different structures of networks produce different results for the same data set.
- Single ANN model cannot produce good generalization for complex data sets having some random errors.
- Generalization capability of a single ANN will be weakened by insufficient training data.
- Many applications are too large for any single ANN to understand in practice.

However, the ensembles will have generalization performance better than single networks only when these are created optimally. The first step of the ensemble design is to select diverse training data sets from the original source data [17]. In many cases, the number of training data samples is limited, so the diverse training subsets are generated from the available limited number of samples. There are many techniques to achieve this, and the most common are: bagging, noise injection, cross-validation, stacking, boosting, and their variants [15]. The second step of the ensemble design is to create different ANN ensemble members. The ensemble members must be both accurate and diverse for good generalization performance [18]. The various ways of achieving these conflicting conditions are: training the ensemble members on different subsets of the data set, different initial random weights for each of the ensemble members, varying the structure of individual members, *i.e.*, different numbers of hidden units and different numbers of hidden layers, using different training algorithms, *e.g.*, the back-propagation, and radial-basis function algorithms [15, 17, 19]. The generalization in the ensemble is also improved when the errors of ensemble members are negatively correlated [20]. The third and last step in ensemble design is to combine the individual ensemble outputs. The conventional strategies used to combine these single ANN results to generate the final output are simple averaging, weighted averaging, majority voting, and ranking [15]. The other methods of combining outputs include the use of evolutionary algorithms to decide the weights assigned to ensemble members [21, 22].

The count of ensemble members also affects the overall performance. There are two alternatives for deciding a number of ensemble members: the first is creating a fixed number of ensemble members which involves rich prior knowledge and experienced human experts, and the second is overproducing ensemble members and then choosing a relevant subset. Two important aspects while deciding the number of members are the search criterion and the search algorithm [23].

The use of multi-objective optimization approaches involving multiple search criteria such as diversity, error rate, and a number of members [19], the multistage ANN ensembles [15], and assignment of complementary weights during the combination of the selected individuals using suitable learning mechanisms [24] are other ways of creating efficient ensembles.

Genetic Algorithm (GA)

GA belongs to a class of stochastic optimization algorithms which is capable of converging to global optimum through iterations. GA is suitable for electromagnetic optimization due to its inherent parallel nature [25]. The genetic operators such as crossover, mutation and reproduction form the basis of GA's selection procedure [26]. These algorithms use a population of finite length strings, called chromosomes, as many initial points, while the gradient-based optimization algorithms start with only one initial point. Thus, this method searches the whole parameter space in parallel, and due to this, the chances of achieving the global optimum increase. The fitness of each individual in the population is evaluated [27], and the members with the higher fitness values are selected to be parents, to which the genetic operators are applied to obtain children. Random mutations are applied to make the population diverse. The parents for the next generation are selected by scoring children and finding the best performers. These steps are iterated until the stopping criterion is met. The commonly used termination criteria are the maximum iteration number, a target fitness value, and a minimum standard deviation criterion [28]. The final global best location gives the required optimized solution. The flow chart shown in Fig. (**2.4**) describes the working of the GA.

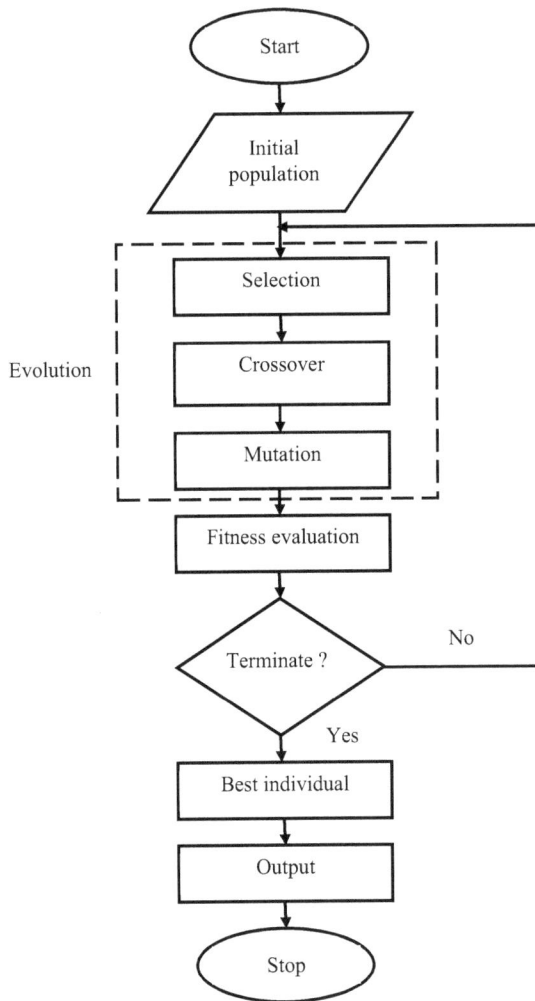

Fig. (2.4). Flow Chart of GA.

Particle Swarm Optimization (PSO) Algorithm

PSO algorithm, a global optimization method, was proposed by Kennedy and Eberhart in 1995 [29]. This optimization technique is very efficient for solving problems in which the optimal answer can be represented in an N-dimensional space as a point or surface. The PSO algorithm is motivated by the collective behavior of birds or swarms called particles. The particles have memory and assist each other to reach the global optima [30]. At the start of the optimization process with PSO, the solution space is defined, which means the variables to be optimized are initialized. The minimum and maximum values of each variable should also be defined [31]. In this algorithm, an initial population, which is a group of particles with arbitrary locations in the N-dimensional solution space, is defined. Each particle has an associated velocity with random direction and magnitude. The position of each member of the population represents a potential solution to the optimization problem [28]. The goodness of these individual locations is evaluated by using a cost function that takes the position co-ordinates as input and returns a single number as fitness score. The cost function is also known as the objective function or the fitness function. The cost function should be chosen critically and accurately because the performance of the PSO algorithm largely depends on it. The functional relationship of the variables to be optimized with the global optima and their relative importance should be demonstrated by the cost function [31]. The solution space and the cost function are specific to the problem to be optimized; however, all other steps are independent of the optimization problem. In each iteration, PSO lets every particle move from a given position to a new one with a velocity calculated by using the best position of the particle (called *pbest*), and the best position of the group (called *gbest*) [32]. During the optimization process, each particle adjusts its velocity to move toward the best solution. The PSO algorithm employs the following two equations for updating the velocity and the position of the particles [31].

$$v_i(t+1) = wv_i(t) + c_1r_1[pbest(t) - x_i(t)] + c_2r_2[gbest(t) - x_i(t)] \quad (2.6)$$

$$x_i(t+1) = x_i(t) + v_i(t+1) \quad (2.7)$$

where v_i is the velocity of the particle, x_i is position of the particle, w is inertial weight, r_1 and r_2 are two random numbers, c_1 and c_2 are parameters which are related to the cognitive and social behavior of the particles. The value of c_1 influences the exploration of the solution space and that of c_2 exploitation of the solution space. The algorithm stops either by obtaining an acceptable target solution or after running for a maximum number of search iterations or any other predefined criterion [33]. The flow chart shown in Fig. (**2.5**) illustrates the sequence of various steps of the PSO algorithm.

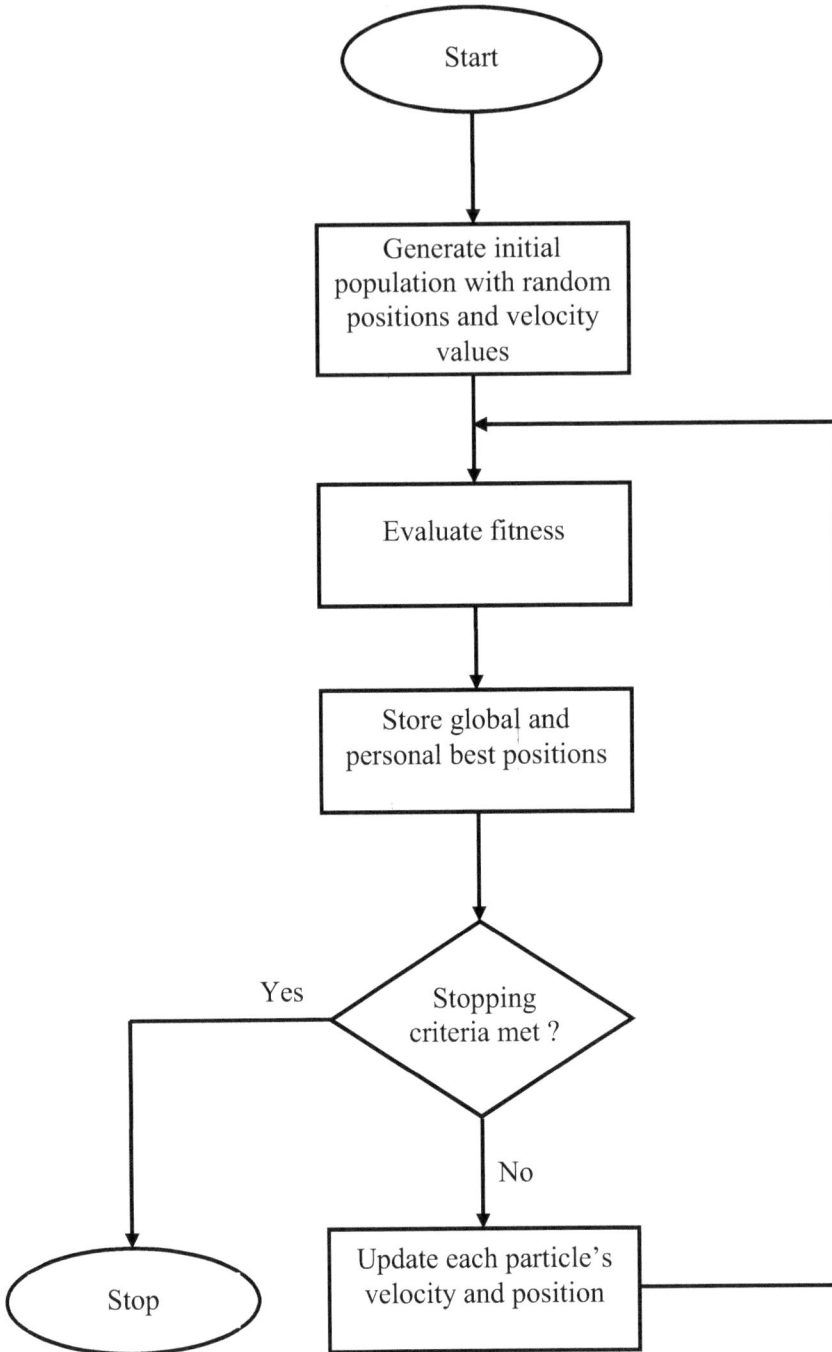

Fig. (2.5). Flow Chart of PSO Algorithm.

Bacterial Foraging Optimization (BFO) Algorithm

The BFO algorithm is a relatively new algorithm proposed by Passino in 2002. This approach is based on the behavior of *E. Coli* bacteria, which is present in the human intestine, for searching the food [34]. This algorithm also, similar to the GA and PSO, starts with a population of bacteria. The operations used to mimic the behaviour of the *E. Coli* bacteria are: chemotaxis (swimming & tumbling), elimination & dispersal, and reproduction [35]. During the chemotaxis step, the bacteria decide whether they should swim, *i.e.*, travel in the current direction, or tumble, *i.e.*, move in a new direction so that they shift to areas which are rich in nutrients and free of noxious substances. The bacteria interchange between these operations, *i.e.*, swim and tumble for the whole lifetime [36, 37]. After a specified number of chemotactic steps, the reproduction process is started and achieved in two stages. In the first stage, fitness scores are assigned to all bacteria present in the search space and ranking is done on the basis of this fitness value. The second stage involves killing off the one-half least healthy bacteria of the population and splitting of each surviving bacterium into two copies of itself placed at the same location [38]. So, the total number of bacteria remains the same. The environment where the bacteria exist may undergo certain steady or unexpected changes which affect the lives and the locations of the bacteria. For example, a considerable increase in the value of temperature can happen, which will wipe out the nearby groups of bacteria. Similarly, there may be changes that cause the shifting of bacteria to new locations. In the BFO algorithm, these effects are replicated in the elimination & dispersal step. This step involves two activities, the first is the elimination which means randomly killing off some bacteria within the population and the second is the dispersal which initializes the bacteria (equal to eliminated bacteria) randomly over the search space so that the total population remains constant [39]. Finally, the algorithm stops when a specified number of iterations of the above operations have occurred. At the point of termination, the algorithm outputs the best fitness value and corresponding design variables [35].

HYBRID BIO-INSPIRED COMPUTING TECHNIQUES

Many hybrid bio-inspired optimization techniques have been proposed in the last decade to enhance the accuracy and computational powers of conventional bio-inspired algorithms. Another reason for designing hybrid algorithms is to reduce the computational times of complex optimization problems. In the field of antennas, hybrid techniques are used to find more efficient solutions for many problems. [40] employed the hybrid of GA and PSO to develop an optimal profiled corrugated horn antenna. Pantoja *et al.* [41] presented a hybrid of GA and space-mapping techniques for antennas in which the space-mapping stage follows the GA approach, and it resulted in improved accuracy and reduced computational cost. An efficient hybrid of GA and BFO is proposed by [42] to solve global optimization problems. Another hybrid algorithm, named the genetical swarm optimization, having properties of GA and PSO algorithms, was developed by Grimaccia *et al.* [43], which is very effective to solve the electromagnetic problems. The PSO and BFO were hybridized by Gollapudi *et al.* [44] to enhance the accuracy of resonant frequency calculation of rectangular microstrip antennas. The accuracy and speed of BFO algorithm were improved [45] by merging the PSO in the chemotaxis step of BFO. The enhancement of global search capabilities of BFO by hybridization with GA has been proposed in [38]. All hybrid algorithms have one or more advantages over the individual constituent algorithms, so these hybrid techniques, along with the growing capability of computers, resulting in more acceptable solutions to the optimization problems.

BIO-INSPIRED COMPUTING TECHNIQUES IN ANTENNAS

ANN Applications in Antennas

The ANNs have found a number of applications in the field of antennas. The applications include the analysis of antennas, synthesis of antennas, parameter estimation, antenna arrays and many more. The applications presented below are classified into different groups depending upon the functions performed by the ANN model.

Microstrip Antenna Design using ANNs

The design of a rectangular patch antenna using ANN is presented by Mishra and Patnaik [46]. They employed ANN for the accurate estimation of the patch length for the given set of other parameters and for different feeding techniques. The performance of RBFNN and different MLPNNs are compared for the design of rectangular patch geometry by Turker *et al.* [47]. Panda *et al.* [26] and Khuntia *et al.* [48] presented the coupling of GA and ANN for the design of a rectangular patch antenna on a thick substrate. A trained ANN is taken as an objective function in GA and the optimized dimensions of the antenna are calculated. It is shown that the results obtained by this method are similar to those obtained by experimental and other methods. A new geometrical methodology based on ANN principles, combined with compact and broadband design principles, is presented by Lebbar *et al.* [49]. Five applications of this methodology resulting in compact-size antennas are presented. The design of an equilateral triangular microstrip antenna using a PSO-driven RBFNN is presented by Chintakindi *et al.* [50]. The neural network weights are replaced by the particle position equation, and the rate of change of the neural network is replaced by the particle velocity equation. A microstrip circular ring antenna is designed using ANN for specified multi-frequency operation by Siakavara [51]. A multilayered dielectric substrate is used for this probe-fed printed slotted ring antenna. An MLPNN model is used to find the width and the position of the slits along with the values of the structural parameters and the feed position of the antenna for desired frequency performance. The parameter estimation of a multi-slotted rectangular microstrip patch antenna using ANN is described in Kumar *et al.* [52]. Different ANN types have been developed, and the results are compared with simulation results. An ANN model for the simultaneous calculation of multiple output design parameters of circular microstrip antennas is presented by Gultekin *et al.* [53]. Different learning algorithms are used to train the networks and it has been concluded that the extended delta-bar-delta algorithm has the best performance. An ANN-based model for the design of a circularly-polarized square microstrip antenna with truncated corners is proposed by Wang *et al.* [54]. The training data sets are prepared by using the empirical formulas of the resonant frequency and Q-factor. The side length of the square and the size of the truncated corners are taken as outputs of ANN model. The synthesis model used three hidden layered networks trained with the Levenberg-Marquardt algorithm. Bose and Gupta [55] have presented an ANN model using a hybrid neural network for the design of aperture-coupled microstrip antennas by combining RBF and a back-propagation algorithm. The hybrid model determines various output parameters such as dimensions of ground plane, dimensions of aperture, dimensions of the radiating element, dimensions of feed, and feed position for desired frequency of operation. The comparison of the results of hybrid model reveals that the hybrid approach is better than the conventional RBF and back-propagation algorithm models in terms of percentage error and execution time. An application of ANN for the design of an inset-fed rectangular microstrip antenna is discussed by Vilovic *et al.* [56]. In this paper, the PSO is used to find the optimum set of weights of ANN. The values of resonant frequency and substrate parameters have been used as inputs. The width & length of the patch and inset feed distance are outputs of the ANN model.

Resonant Frequency Calculation using ANNs

The use of ANNs to calculate the resonant frequency of different rectangular microstrip patch antennas is presented by Devi *et al.* [57], Khuntia *et al.* [58], and Pattnaik *et al.* [59]. These papers employ GA for the design of ANN. The GA is used to select the parameters of ANN which leads to decrease in the time required for training of neural network. A fast technique using neuro-fuzzy networks to evaluate the resonant frequency of microstrip antennas is proposed by Angiulli and Versaci [60]. Guney and Sarikaya [61] proposed an Adaptive Network based Fuzzy Inference System (ANFIS) to calculate simultaneously the resonant frequencies of various microstrip antennas of various geometries. The results of the proposed hybrid method for the rectangular, circular, and triangular antennas are compared with the experimental results and are found in very good agreement. The application of ANN for the calculation of resonant frequency of electrically thick and thin circular is proposed by Gangwar *et al.* [62]. The ANN results are sufficiently accurate as compared to theoretical and measured results. An ANN ensemble is used by [22] for modeling the resonant frequency of rectangular microstrip antenna. In an ANN ensemble, several ANNs are trained and then the results are combined by following certain rules. This improves the generalization ability of the ANN system. Yu-Bo *et al.* [22] used the decimal PSO algorithm and binary PSO algorithm to select ANNs to construct ANN ensemble. The inputs to the ensemble model are antenna dimensions and substrate parameters. The output is the corresponding resonant frequency. The ANN ensemble method provided better results than simple ANN methods. A multilayer ANN structure combined with EM knowledge trained with spectral domain approach responses for the calculation of the resonant frequency of the rectangular microstrip antenna with and without air gap printed on isotropic substrate is proposed by Tighilt *et al.* [63]. The EM knowledge is used to preprocess the ANN model inputs to reduce the training time and to increase the accuracy. A multilayer feed-forward ANN model for estimating the dual operating frequencies of a shorting pin-loaded dual-band equilateral triangular microstrip antenna is proposed by Can *et al.* [64]. The inputs of the model are the side length, thickness, permittivity, and shorting pin position ratio. The comparison of results with theoretical and measured results has shown the considerable improvement achieved over the recent studies.

ANNs for Other Antenna Parameter Calculations

The calculation of the radiation efficiency of a rectangular microstrip patch antenna with the use of ANN trained using the back-propagation approach is discussed by Hettak and Delisle [65]. The method proposed by them is valid for substrates with relative permittivities between 1 and 12.8 and for the complete range of thicknesses normally used for antennas. Neog *et al.* [66] presented the use of a tunnel-based ANN model for the parameter calculation of a wideband microstrip antenna. The proposed tunneling technique used in the traditional gradient descent back-propagation algorithm takes less computational time as compared to the back-propagation model while producing accurate results. A method based on ANFIS is presented by Guney and Sarikaya [67] to compute the bandwidth of a rectangular microstrip antenna. They used different optimization algorithms to determine optimally the design parameters of the ANFIS and found that the ANFIS model trained by the least-squares algorithm generates the best results. The use of finite impulse response ANN for speeding up Finite-Difference Time-Domain (FDTD) calculations is presented by Panda *et al.* [68]. This neuro FDTD technique is applied to calculate the input impedance of a coaxially fed stacked microstrip patch antenna and it has been found that the same result with less number of iterations as compared to convention FDTD can be achieved. An RBFNN model is proposed by Vilovic and Burum [69] for the estimation of the feed point of a circular microstrip antenna. The radial distances from the center of the patch are taken as input and the corresponding input impedance is taken as the output of the model. An ANN-based synthesis model is presented by Khan *et al.* [70] for estimating the slot-size on the radiating patch and air-gap inserted between the ground plane and the dielectric substrate for obtaining the improved performance of a dual band rectangular microstrip antenna. 10-dimensional input consisting of dual resonance frequency, dual-frequency gain, dual-frequency directivity, dual-frequency antenna efficiency, and dual-frequency radiation efficiency parameters is used for the design. An RBFNN model is proposed by Aneesh *et al.* [71] for the bandwidth analysis of a slot-loaded microstrip line feed patch antenna. The slot-loaded antenna has triple-band frequency performance, and the ANN model estimates the bandwidth in all bands accurately for different slot lengths and widths.

Other Antenna Related Applications of ANNs

An ANN model of a MEMS-switched frequency-reconfigurable antenna is presented by Patnaik *et al.* [72]. The developed ANN finds the location of the operational frequency bands for any combination of switches connecting different radiating elements. This technique significantly reduces the computational complexities involved in the numerical modeling of reconfigurable antennas. A review of applications of ANNs for different types of antennas, such as microstrip antennas, CPW patch antennas, wideband antennas, and multi-band antennas, is provided by Patnaik *et al.* [73]. Some examples of ANN applications in smart antennas, reconfigurable antennas, and antenna arrays are also presented. A new type of ANN, the Synthesis-ANN, which uses a hetero-associative memory and extends a subspace containing training data to solution subspaces by random variations of inputs and outputs, is proposed by Delgado *et al.* [74]. The proposed model is applied to the optimization of a printed dipole antenna with an integrated balun, and it is seen that the ANN arrived in the neighborhood of solution subspaces within a few iterations. The neural-genetic optimization is used by Dubrovka and Vasylenko [75] for the optimization of a "bow-tie" type planar antenna for UWB performance in the frequency range 3.1–10.6 GHz. The desired operation is achieved by suitably adjusting the radiating contour profile of the conventional triangular taper of the bow-tie antenna. Mudroch *et al.* [76] proposed an approach for the optimization of different UWB dipole antennas using an algorithm tuned using ANN. The optimization process searches for the dipole shape, which meets a good matching and a minimal distortion of radiated impulses. The synthesis of UWB planar antennas utilizing a model built upon ANN and their inversion by GA is presented by Dubrovka and Vasylenko [77]. A dipole antenna and a slot profile of the Vivaldi antenna are synthesized using the proposed hybrid algorithm which is 5 times faster than GA and 1.5–2 times faster than PSO algorithm. The design and optimization of an annular ring dielectric resonator antenna using an ANN approach is presented by Lucci *et al.* [78]. A harmonic tuning technique for a slot-coupled annular dielectric resonator antenna is proposed so that the antenna operates in the *C* band of the frequency spectrum. The mutual coupling between the radiating elements of a rectangular MIMO antenna for a specified range of separation between the antennas and for various frequencies is studied by [79] using ANNs. The proposed antenna is developed on a flexible substrate, which can be used for wearable applications. Different ANN algorithms are used to train the neural structure, and the comparison between them shows that the quasi-Newton and quasi-Newton multi-layer perceptron algorithms are better in terms of various performance measures. Gunes *et al.* [80] employed an MLPNN model based on the 3-D CST microwave studio software in both design and analysis of the Minkowski reflectarray antenna. The MLPNN model is also used to observe the effect of the feed movement along the focal length on the gain-bandwidth and the radiation pattern. All steps of designing the MLPNN model and its utilization in the design and analysis of a Minkowski reflectarray antenna are explained as a general method, so, the proposed scheme is applicable to different types of antennas.

Applications of Bio-Inspired Optimization Techniques in Antennas

The optimization algorithms have found a number of applications in the area of antenna design and optimization in the past two decades. The extracts of some important papers on the use of optimization algorithms in antenna design are presented as follows:

Several applications of GA-based methods for antennas have been described by [25]. These include single-element antenna applications like optimization of wire antennas, reflector antennas, and Yagi antennas. Also antenna array applications like reducing the side-lobes of an array, reducing the scattering of strip arrays, arrays with specified null locations and a specific pattern shape arrays have been described. The use of GA to determine the optimized dimensions of microstrip antenna of rectangular shape has been presented in Pattnaik *et al.* [81]. The GA model inputs are the desired frequency of operation and substrate parameters; and the optimal length and width values are provided as outputs. The electrically thin antennas are selected for the design. The design of a patch antenna on the thick substrate by combining the GA and ANN is described [26]. A trained ANN is used as a cost function in GA for finding optimal results. The optimized values calculated by this approach are nearer to experimental dimensions. The design of multi-band patch antennas using the cavity-model-based simulation tool and the

GA is presented by Ozgun *et al.* [27]. The antennas described in their work have employed a number of slots in the patch or several shorting strips between the ground plane and the patch. In the first step, the variation in the input impedance due to the slots and shorting strips is analyzed. Then in the second step, GA is used to determine the optimal locations of slots and shorting strips, so that the antennas behave as per the requirements over the selected resonant bands. The synthesis of a complex phased array to achieve a desired far-field sidelobe value using GA and PSO, is described by Boeringer and Werner [28]. The amplitude-only, phase-only and complex weighting arrays are designed with excellent performance using PSO and GA.

Robinson and Rahmat-Samii [31] presented a comprehensive description of the PSO algorithm and explained its suitability in electromagnetic optimizations. The optimization of a profiled corrugated horn antenna is also described in detail. Kerkhoff *et al.* [82] designed and analyzed planar monopole antennas using GA. The impedance bandwidth of two shapes, the bow-tie and reverse bow-tie, are optimized by employing GA. One outcome of this paper is that the bandwidth of the reverse bow-tie is greater than the bow-tie, and it has a very small relative size. Then the randomly shaped monopole antennas are designed using GA, and these antennas have shown enhanced performances and less sizes in comparison to the reverse bow-tie. The design of a self-structuring antenna (antenna competent of assembling itself into numerous configurations) using the simulated annealing, ant-colony optimization, and GA is presented by Coleman *et al.* [83]. The realization of each method is discussed, and the results of all techniques are compared to a random search.

Jin and Rahmat-Samii [84] presented a methodology based on a combination of PSO and FDTD for the design of patch antennas having multiband and wide-band performance. The geometric parameters of the antenna which are to be optimized are extracted by PSO, and FDTD evaluates a cost function that returns a fitness value which is used to rank the possible designs as per their performance. Two examples, first, the design of rectangular patch antennas and second, the design of E-shaped patch antennas, are investigated in this paper. The dimensions of a rectangular microstrip antenna are calculated by Khuntia *et al.* [48] using a hybrid of ANN and GA. The trained ANN is employed as the cost function of a binary-coded GA. The comparison of GA and PSO for the multi-objective optimization of wire multi-band antennas is described by Lukeš and Raida [85]. Antennas are optimized to reach the desired matching, demonstrate the omni-directional constant gain, and have reasonable polarization purity. Liu [33] proposed a PSO-based approach to design a multiband monopole antenna having a CPW feed. Impedance matching is obtained for multi-resonant mode by introducing slits into the feeding line. The design parameters of the antenna for desired multiband performance are obtained by using a PSO algorithm in conjunction with the method of moments (MoM).

The application of GA for the optimization of electrically small wire antennas is reported by Choo *et al.* [86]. A multi-objective GA is implemented to simultaneously optimize bandwidth, efficiency and antenna size. Efficient optimization is achieved by involving three chromosomes with a two-point crossover scheme and a geometrical filter. Soontornpipit *et al.* [87] described the use of GA for the designing of a waffle-type antenna appropriate for operation in MICS band (402 – 405 MHz). The design is optimized to get S_{11} less than −15 dB throughout the whole band. This is achieved by simulating the antenna with FDTD and improving the design using a GA. Misra *et al.* [88] used GA to design an asymmetric V-dipole antenna and its three-element array. The directivity is taken as the optimization parameter for the V-dipole, and that for the array are input impedance and directivity. The optimization of antenna for UWB applications using GA is proposed by Telzhensky and Leviatan [89]. The time-domain characteristics are used to optimize a variant of the volcano smoke antenna. The aim of the optimization procedure is to find an antenna with low VSWR and low-dispersion. Dual-band implantable antennas optimized for MICS and ISM bands using a PSO and HFSS implementation have been described by Hood and Topsakal [90]. The E-shaped patch and many other geometries of antennas are discussed.

A hybrid of GA and the space-mapping tool is proposed by Pantoja *et al.* [41] for the estimation of optimal lengths and feed locations for an array of microstrip antennas. The GA used coarse models, and the space mapping improved the result obtained in the GA step. The computational cost of the hybrid approach is less as compared to the single application of GA with a fine-model simulator. The design of UWB planar antenna using a hybrid algorithm of two-dimensional GA and FDTD is presented by Ding *et al.* [91]. The Best-Mate-Worst strategy is used for the mating of chromosomes to avoid premature convergence. The design of a small UWB antenna which operates over a resonant band from 3.1 GHz to 12 GHz is proposed. Pérez and Basterrechea [30] presented a comparison between various optimization algorithms like simulated annealing, GA, PSO and their several modified forms for the reconstruction of a far-field radiation pattern of an antenna from planar near-field data. PSO algorithm in which the swarm is updated asynchronously and has a global topology and the simulated annealing scheme with search step perturbation is concluded as the best algorithms on the basis of accuracy, simplicity and computational load. PSO has been employed to estimate the resonant frequency and feed location of rectangular microstrip antenna by Chintakindi *et al.* [92]. The values provided by the PSO algorithm are very near to experimental results and are achieved with a reduced computational time. Jin and

Rahmat-Samii [93] presented single-objective and multi-objective optimizations of nonuniform and thinned antenna arrays using real number PSO and binary PSO. Real-number PSO achieved a better value of peak side-lobe level by optimizing the positions of the elements in a non-uniform array. The ON and OFF state of elements of a thinned array is decided using a binary PSO. Multi-objective cases are considered to find the non-dominated solutions on the Pareto front to optimize the other design parameters than the peak side-lobe level. A hybrid algorithm called genetical swarm optimization combining the properties of PSO and GA is presented by Grimaccia *et al.* [43] for the optimization of a linear array. A key characteristic of the proposed optimizer is that the algorithm dynamically sets a driving parameter, called the hybridization coefficient, in order to adapt itself to any specific problem.

Rattan *et al.* [94] presented an application of PSO for linear array designing using half-wave parallel dipoles. The side-lobes are suppressed, and null assignment in desired directions is achieved by arranging the half-wave parallel dipoles as a non-uniform linear array. A germ swarm, a hybrid optimization algorithm, is presented by Gollapudi *et al.* [44] to compute the frequency of operation of a rectangular patch antenna. An improvement in the correctness with less processing time is reported. An improved adaptive BFO algorithm is proposed by Datta *et al.* [95] to optimize an array of antenna elements. The objective of the optimization is to maximize the array factor in a specific direction and null placements in desired orientations. It is described that the presented method takes less time and increases the accuracy of the optimized solution. The side-lobe level of a circular antenna array having a non-uniform configuration is reduced by optimization using PSO by Shihab *et al.* [96]. The spacing of elements of an array and the optimal weights are decided by employing the PSO so that the side-lobe level of the radiation pattern is reduced while maintaining a desired major lobe beamwidth. The synthesis of antenna arrays using GA, memetic algorithm and Tabu search algorithm is discussed by Cengiz and Tokat [97]. Three examples of antenna array design are presented to compare the efficiency of the algorithms. The memetic algorithm finds the most convenient results due to its local search capability but at the cost of increased time of iterations. Chintakindi *et al.* [98] explored the effectiveness of PSO in the estimation of the frequency of operation of a patch antenna having an equilateral triangular shape, and it has been found that the PSO can be used for such applications. The optimal values obtained using the PSO are very close to the experimental and GA results. The use of BFO for the calculation of frequency of operation and the feed location of rectangular patch antenna is explained by Gollapudi *et al.* [35]. The accurate results are achieved with reduced computational time. The PSO has been used to design a multiband microstrip antenna, an antenna array, and an artificial ground plane by Jin and Rahmat-Samii [99]. The simulated and measured results of optimal designs are proposed, which confirm the effectiveness of PSO in developing beneficial and realistic solutions. The use of PSO for the design of an implantable antenna with a dual resonant band is presented by Karacolak *et al.* [100]. The gels having electrical properties similar to that of human skin are used for validating the developed antenna, and the optimal design of the antenna is fabricated and experimented with in the presence of the developed gel.

The BFO and PSO algorithms are compared by [36] for an antenna-array optimization problem. The PSO produced superior solutions for null placement than the BFO, however, BFO performed better in side-lobe suppression. Panduro *et al.* [101] presented a comparison of GA, PSO and the differential evolution (DE) optimization algorithms for developing circular antenna arrays with scannable properties. An intelligent BFO technique obtained by hybridizing BFO with PSO to reduce the convergence time and enhance the accuracy is proposed by Gollapudi *et al.* [102] and this technique has been applied to calculate the frequency of operation of the rectangular microstrip antenna. A new design model called the semi-automatic design of antennas using an adaptive improved comprehensive learning PSO is introduced by Wu *et al.* [103]. This model contains two steps: in the first step, a digital configuration of an antenna is obtained, and in the second step, its dimensions are optimized further according to the current distribution of the digital one. The design of a small multiband printed monopole antenna is presented as an example. Mahmoud [104] optimized a bow-tie antenna for a 2.45 GHz band application using a hybrid of PSO and BFO called the bacterial swarm optimization algorithm along with a simple algorithm called Nelder-Mead. The bacterial swarm optimization - Nelder-Mead method has yielded solutions superior to that of BFO, bacterial swarm optimization and BFO - Nelder-Mead algorithms.

Tseng and Han [105] obtained slot antennas having circular polarization, broadband and multiband properties using a new genetic local search algorithm. The proposed design technique is validated by developing antennas with various slot geometries: the elliptical, the triangular, and the square. The use of a hybrid real-binary PSO technique in the design of a non-uniform antenna array and an antenna with dual-band characteristics for handset applications is discussed by Jin and Rahmat-Samii [106]. The usefulness of the hybrid technique in obtaining solutions with enhanced performance is confirmed by simulation and measurement results. The use of BFO for resonant frequency optimization of a patch antenna of circular shape having air gaps is described in Sharma and Kanaujia [107]. The BFO results are compared with the experimental results by changing diameter of patch, height of air-gap, and mode of resonance and are found to be in good agreement. Yeung *et al.* [108] provided an overview of the application of GA, PSO, and space mapping algorithms for antennas and RF circuit designs. Two design examples, the design of a U-slot microstrip patch antenna and the design of a microstrip pass-band filter, are also discussed. Anjitha and Kumar [109] optimized a Zig-Zag antenna using PSO algorithm. The segment length and vertex angle of the Zig-Zag antenna are optimized to maximize the gain. In conventional Zig-Zag design, constant values of segment dimension and vertex angle are used. The simulated results illustrated that the antenna developed by the PSO algorithm produced superior gain values than the conventional Zig-Zag antenna. Silva *et al.* [110] applied a self-organizing multi-objective GA to a ring monopole microstrip antenna with a slit in the ground plane for UWB applications. Bandwidth, return loss and central frequency deviation are considered as three objectives simultaneously. These objectives are modeled as dependent on two variables, *i.e.*, the slit dimensions, and are used in a single weighted objective function. The weights are calculated by an adjuster GA that uses another GA to estimate each combination of possible weights. Rahmat-Samii *et al.* [111] described the main features, terminology, various boundary conditions and engineering constraints of PSO. This paper also discusses the implementation of PSO algorithm with real and binary parameters. The application of PSO for two different antenna designs: a multiband handset antenna and an E-shaped patch antenna for circularly polarized applications, are presented. The hybrid PSO-MoM program implemented for the multiband handset antenna achieved a small-size antenna design with dual-band performance. The E-shaped patch design with a low-profile antenna with wideband circularly polarized characteristics is achieved by using a PSO – Finite Element Method program.

Minasian and Bird [112] presented a design of a microstrip antenna loaded with parasitic patches for WLAN systems. The parasitic patches are electromagnetically coupled with a co-axial probe-fed rectangular microstrip antenna. The location and orientations of the patch are optimized by a PSO algorithm so that the antenna resonates over 5 GHz to 6 GHz frequency range. The designed antenna has an omni-directional radiation pattern with sufficient gain for WLAN applications. Li *et al.* [113] improved the performance of the traditional PSO algorithm by a neighborhood-redispatch method and named the new, improved algorithm as neighborhood-redispatch-PSO algorithm. The efficiency of the proposed neighborhood-redispatch-PSO algorithm is first validated by applying on benchmark functions and then it is used to design a CPW-fed microstrip antenna. The optimal values of nine design variables are calculated so that the antenna has a resonant band over UWB frequency range with a stop band from 5.15 GHz – 5.825 GHz to evade possible interferences. Silva and Martins [114] described the use of a machine learning approach to optimize a microstrip antenna for UWB applications. Three different objective functions representing bandwidth, return loss, and central frequency deviation are developed using the machine learning approach from the experimental data. This multi-objective optimization problem is converted to a single objective by combing the three objective functions using suitable weights. The design variables are length and width of a slit used in the ground plane to achieve the desired performance. Cismasu and Gustafsson [115] presented an approach to compute Q factors of antennas using single frequency simulation data. The method is based on the values of energies stored in electric and magnetic fields excited by the radiating structure. The integration of the proposed method with GA is discussed to optimize the antenna bandwidth with reduced processing time. The performance comparison of the DE, GA, PSO, and their variants is evaluated by Deb *et al.* [116] for the optimization of microstrip antennas over a desired frequency range. The design objectives are impedance matching for obtaining circular polarization and high gain. Three different feed mechanisms, namely co-axial, microstrip-line, and aperture coupled, are evaluated. Four different objective functions are developed and then combined into a single function with dynamic weights allocated using a fuzzy approach. It has been found that the DE-based approaches have better performance for the selected antennas. Manh *et al.* [117] described the use of ANN as an objective function of a PSO algorithm for antenna optimization. A proximity-coupled multilayer dual rectangular ring antenna is optimized by the proposed approach. A new approach of using separated ANNs as surrogate models is used to achieve better time convergence and accuracy. Koppisetti *et al.* [118] designed an adaptive antenna system using GA and PSO techniques. GA is used to find the solutions, which are then refined by applying the PSO algorithm. The performance of the method is evaluated by conducting system-level simulations. The improved coverage and system throughput is achieved as compared to the fixed antenna systems. Also, the presented method is computationally efficient and simple to implement.

Limitations of the Existing Bio-Inspired Computing Techniques

The main limitations of the existing bio-inspired computing techniques are as follows:

- Limited local and global search capabilities.
- A large number of iterations required for achieving global optima.
- Sequential calculations.
- Absence of suitable objective functions to optimize irregular antenna shapes.
- Limited generalization leads to the need for an efficient hybrid algorithm.

APPLICATIONS OF THE BIO-INSPIRED OPTIMIZATION TECHNIQUES IN FRACTAL ANTENNAS

Bio-inspired optimization techniques have found a number of applications for analyzing and designing the fractal antennas. This section discusses the extracts from some important papers.

The initial work in the domain of optimization of the fractal antennas is proposed by Werner *et al.* [119]. An IFS algorithm is employed to generate fractal antenna geometries and the VSWR requirements are satisfied by using two inductive-capacitive loads. GA-based approach is employed to design a dual-band fractal antenna by the optimization of the shape of fractal antenna, position of the loads, inductive & capacitive values of loads, and overall length of the antenna element. A sensitivity analysis on the load component values of these antennas suggests that the performance varies with the load component values. Several optimization approaches have been proposed by Werner *et al.* [120], which lead to antenna designs with considerably reduced load sensitivity. Pantoja *et al.* [121] developed Koch-like pre-fractal wire antennas using a multi-objective GA by optimizing their efficiency and bandwidth while reducing their frequency of operation. The initial GA population consists of the pre-fractal antenna elements, and it is generated using an IFS algorithm. Three different cost functions are used to determine the fitness of individuals in the population and the algorithm uses these fitness scores to reach an optimized solution. The multi-objective GA procedure optimizes all three fitness functions at the same time and provides a number of optimized solutions which are used to choose the antenna elements that meet the specified optimization objectives. Werner and Werner [122] demonstrated that antennas can be optimized *via* GA to achieve superior performance characteristics (*e.g.*, input impedance, VSWR, and gain) when placed in close proximity to a perfect magnetic conductor ground plane. This technique is used to develop a miniaturized fractal antenna in dipole configuration. The optimization of a crown square fractal microstrip antenna using GA to get a minimum axial ratio is described by Dehkhoda and Tavakoli [123]. The optimization parameters are the ratio of the two sides of the antenna and the diagonal feed location. Fractal tile geometries-based patch antennas are introduced by Spence and Werner [124] to design single-feed microstrip antennas having gain in broadside direction greater than 12 dB. The use of GA has been proposed to optimize the performance of these antennas.

The design of a Koch-like fractal miniaturized monopole antenna for ISM-band application is described by Azaro *et al.* [125]. The proposed antenna is designed by optimizing the fractal geometry and the segment widths through a PSO algorithm. The same approach is followed by Azaro *et al.* [126] for the synthesis of a reduced-size pre-fractal monopole antenna suitable for WiMax band frequencies from 3.4 GHz to 3.6 GHz. The optimization of closed-loop fractal shapes of Sierpinski in the form of Delta wire, Y Wire, and Koch is proposed by Pantoja *et al.* [127]. A multi-objective GA is used to develop new thin wire antennas which are electrically small and have improved characteristics than various pre-fractal Sierpinski antennas. Polpasee *et al.* [128] described the synthesis method of fractal geometry with maximized directivity by using GA. In this method, the position of the element is strategically combined with numerical technique. This application has shown that the GA can be used for developing the square-planar fractal array with improved directivity. Another synthesis procedure for developing compact fractal antennas suitable for various applications using PSO is presented by Franceschini *et al.* [129]. A trapezoidal fractal generator is used to reduce the overall length of the design and to obtain the desired VSWR values.

Ghatak *et al.* [130] described the optimization of a CPW-fed Sierpinski carpet fractal antenna using a real-coded GA. The second-order Sierpinski carpet antenna is developed with wide impedance bandwidth. The gap between the ground plane and middle strip, the distance between the ground plane and radiating carpet, and the size of the CPW central conductor are taken as design variables. The real coded GA is combined with IE3D electromagnetic simulation software, to obtain the optimized values. The same approach is followed by Ghatak *et al.* [131] for the design of the Sierpinski gasket microstrip antenna. A PSO-based approach for the development of a multi-band fractal antenna has been proposed by Azaro *et al.* [132]. The dimensions of a Sierpinski-like fractal shape are optimized using PSO and IE3D electromagnetic simulation software to attain the specifications of the Wi-Fi and GPS frequency bands. Another example of a triple-band fractal antenna is described by Azaro *et al.* [133]. The use of this technique for designing a quad-band fractal antenna with a similar shape is presented in Rocca *et al.* [134]. Vidal and Raida [135] discussed the synthesis of a self-affined Sierpinski monopole antenna operating at three different frequencies. The self-affined Sierpinski has different scale factors for different directions. Antennas are optimized by GA and PSO separately in order to operate in prescribed frequency bands, and it is found that impedance matching is slightly better in the case of GA as compared to PSO. Azaro *et al.* [136] described the design of a compact triple-band patch antenna. The base antenna shape has been named a hybrid prefractal shape, and it is generated by joining two diverse fractal geometries, a Sierpinski-like and a Meander-like structure. The antenna has been optimized using PSO combined with a hybrid pre-fractal geometry generator and IE3D electromagnetic simulator to achieve desired results. It is observed by analyzing different "intermediate" geometries that the merging of the Meander-like geometry with the Sierpinski-like prefractal shape results in a resonant frequency drifting towards the lower end of the frequency scale. Lizzi *et al.* [137] proposed a triple-band fractal antenna optimized for different frequency bands. The base geometry is a Sierpinski pre-fractal geometry with the first three iterations. The ultimate antenna geometry is generated by using an iterative technique that integrates an MoM-based electromagnetic simulator and PSO to arrive at a geometry that meets the required specifications in terms of size and impedance characteristics. Hazdra *et al.* [138] proposed a set of software tools for the design and optimization of fractal patches described by the IFS. The presented approach makes use of a fast cavity model and PSO. The optimization objective is to minimize the fundamental resonant frequency of the fractal patch while preserving the maximum patch dimension. Hieu *et al.* [139] described the use of PSO to design a triple-band Sierpinski gasket-based fractal antenna. The fractal structure of Sierpinski gasket, along with a matching line, is selected as the base shape. The design variables chosen are dimensions of a matching line, angle and side dimensions of the triangle patch, scale of triangle slot, and width.

Oliveira *et al.* [140] described the use of GA for the optimization of fractal antennas based on a triangular Koch shape. The inset-feed is used and it is optimized to minimize the S_{11} value. The quasi-fractal antennas considered in their paper are designed using Koch curves and the excitation of structures is done using a microstrip line. Capek *et al.* [141] described a MATLAB-based code for the easy creation of planar IFS fractals. The generated fractal patches are analyzed by the cavity model, or by characteristic modes. The IFS parameters are then optimized using the PSO optimizer in order to minimize the fundamental resonant frequency. The proposed technique is explained for a second iteration structure called as fractal clover leaf antenna. The L-probe mechanism is used to feed the proposed antenna to achieve broadband operation. A similar design technique for microstrip fractal patch antennas is also proposed by Capek and Hazdra [142]. Adelpour *et al.* [143] presented a modified Koch fractal configuration-based dual-frequency microstrip antenna. Some perturbation to the conventional geometry of Koch fractal shape antenna for achieving new configuration with desired performance is investigated using an evolutionary method based on real coded GA. Anuradha *et al.* [144] employed ANN and PSO for the design of fractal antennas for desired frequencies. The ANN is trained to estimate resonant frequencies of the fractal antennas from the given dimensions of the antennas, and this trained ANN is used by PSO to design the optimal geometry for the desired resonant frequencies. Two examples, *i.e.*, Sierpinski gasket antenna and Koch monopole antenna, are developed using this approach. The same approach is proposed by Ouedraogo *et al.* [145] for designing Sierpinski carpet-based customized fractal frequency selective surfaces for desired frequency characteristics. The optimal design of a fractal monopole antenna having dual-band operation appropriate for LTE communications is presented by Lizzi and

Massa [146]. The base shape for the proposed fractal antenna is perturbed Sierpinski fractal geometry, and PSO is used to determine various geometrical parameters so that the optimized antenna exhibits an excellent matching characteristic for the required bands.

The use of ANFIS for the optimization of impedance bandwidth and radiation pattern of a fractal antenna is proposed by Krishna *et al.* [147]. The ANFIS results for a concentric nano-arm fractal are compared with the results of HFSS software and good matching is reported. Weng and Hung [148] proposed an H-shaped new fractal antenna. The antenna has multiband performance and PSO algorithm is used to optimize the presented antenna for 2.45 GHz and 5.5 GHz WLAN applications. The performance of the antenna is validated experimentally. Ghatak *et al.* [149] optimized the Haferman carpet antenna array using DE and PSO algorithms. The optimization algorithms are used to find optimal excitations, element spacing, and number of elements. The performance is compared with the conventional Sierpinski carpet array and a lower peak side low level with reduced radiating element is achieved with Haferman carpet fractal array.

CONCLUSION

This chapter highlights the importance of bio-inspired computing techniques for antennas. The working principles of bio-inspired computing techniques, namely ANN, GA, PSO and BFO, are described. The applications of feed-forward ANNs in microstrip antennas are discussed, and it is found that ANNs are used for different parameter estimation applications. A number of papers related to the estimation of antenna dimensions using ANNs are reviewed, and their extracts are presented. The other ANN applications, like resonant frequency calculations, bandwidth estimation *etc.*, are described to present the suitability of ANNs in antenna design. Flow-charts of GA, PSO and BFO are shown in this chapter to explain the working of these algorithms. The importance of hybrid optimization techniques is pointed out. Various applications of GA, PSO, BFO and their variants are discussed for antennas. It has been observed from published papers that these algorithms are used for calculating the optimal dimensions of antennas for desired frequencies, computing resonant frequency for given dimensions, and to estimate several antenna characteristics like S_{11}, bandwidth, efficiency, *etc.* The size reduction of antennas is also achieved by these optimization algorithms. The use of ANNs as the objective function is also reported by some researchers as an effective replacement of mathematical expressions relating to input and output variables. The performance comparison of different techniques is also discussed for different antenna applications. The applications of bio-inspired optimization techniques for fractal antennas are reviewed in this chapter, and it has been observed that these optimization techniques are very suitable for the design optimization of fractal antennas. The fractal geometries are complex geometries as compared to simple Euclidian geometries and have a relatively greater number of design variables. The optimal values of design variables are required to achieve the design objectives. Optimization techniques have been used by different researchers to solve this type of problem. These techniques are also employed to tune fractal antennas for certain applications like ISM band applications *etc*. The objective functions of the optimization techniques for fractal antennas are developed empirically, or the electromagnetic simulators are used as the objective function. A few applications of using ANN as an objective function for fractal antenna optimization are also discussed.

DISCLOSURE

Part of this article has previously been published in the following articles:

• B. S. Dhaliwal and S. S. Pattnaik, "Artificial neural network analysis of Sierpinski gasket fractal antenna: A low-cost alternative to experimentation," *Adv. Artif. Neural Syst.*, vol. 2013, pp. 1–7, 2013.

• B. S. Dhaliwal and S. S. Pattnaik, "Performance comparison of bio-inspired optimization algorithms for Sierpinski gasket fractal antenna design," *Neural Comput. Appl.*, vol. 27, no. 3, pp. 585–592, 2016.

• B. S. Dhaliwal and S. S. Pattnaik, "Development of PSO-ANN ensemble hybrid algorithm and its application in compact crown circular fractal patch antenna design," *Wirel. Pers. Commun.*, vol. 96, no. 1, pp. 135–152, 2017.

REFERENCES

[1]   S.N. Sivanandam, and S.N. Deepa, *Principles of soft computing*, Wiley, 2nd ed., India, 2011.

[2]   Q.-J. Zhang, K.C. Gupta, and V.K. Devabhaktuni, "Artificial neural networks for RF and microwave design-from theory to practice", *IEEE Trans. Microw. Theory Tech.*, vol. 51, no. 4, pp. 1339-1350, 2003.
        [http://dx.doi.org/10.1109/TMTT.2003.809179]

[3]   B. Choudhury, S. Thomas, and R.M. Jha, "Implementation of soft computing optimization techniques in antenna", *IEEE Antennas Propag. Mag.*, vol. 57, no. 6, pp. 122-131, 2015.
        [http://dx.doi.org/10.1109/MAP.2015.2439612]

[4]   A. Jain, and A. Kumar, "An evaluation of artificial neural network technique for the determination of infiltration model parameters", *Appl. Soft Comput.*, vol. 6, no. 3, pp. 272-282, 2006.
        [http://dx.doi.org/10.1016/j.asoc.2004.12.007]

[5]   S. Shrivastava, and M.P. Singh, "Performance evaluation of feed-forward neural network with soft computing techniques for hand written English alphabets", *Appl. Soft Comput.*, vol. 11, no. 1, pp. 1156-1182, 2011.
        [http://dx.doi.org/10.1016/j.asoc.2010.02.015]

[6]   B.S. Dhaliwal, and S.S. Pattnaik, "Artificial neural network analysis of Sierpinski gasket fractal antenna: A low cost alternative to experimentation", *Adv. Artif. Neural Syst.*, vol. 2013, pp. 1-7, 2013.

[http://dx.doi.org/10.1155/2013/560969]

[7]   P. Singh, and M.C. Deo, "Suitability of different neural networks in daily flow forecasting", *Appl. Soft Comput.,* vol. 7, no. 3, pp. 968-978, 2007.
[http://dx.doi.org/10.1016/j.asoc.2006.05.003]

[8]   H. Sarimveis, P. Doganis, and A. Alexandridis, "A classification technique based on radial basis function neural networks", *Adv. Eng. Softw.,* vol. 37, no. 4, pp. 218-221, 2006.
[http://dx.doi.org/10.1016/j.advengsoft.2005.07.005]

[9]   D.F. Specht, "A general regression neural network", *IEEE Trans. Neural Netw.,* vol. 2, no. 6, pp. 568-576, 1991.
[http://dx.doi.org/10.1109/72.97934] [PMID: 18282872]

[10]   D. Tomandl, and A. Schober, "A modified general regression neural network (MGRNN) with new, efficient training algorithms as a robust 'black box'-tool for data analysis", *Neural Netw.,* vol. 14, no. 8, pp. 1023-1034, 2001.
[http://dx.doi.org/10.1016/S0893-6080(01)00051-X] [PMID: 11681748]

[11]   K. Nose-Filho, A.D.P. Lotufo, and C.R. Minussi, "Short-term multinodal load forecasting using a modified general regression neural network", *IEEE Trans. Power Deliv.,* vol. 26, no. 4, pp. 2862-2869, 2011.
[http://dx.doi.org/10.1109/TPWRD.2011.2166566]

[12]   L.K. Hansen, and P. Salamon, "Neural network ensembles", *IEEE Trans. Pattern Anal. Mach. Intell.,* vol. 12, no. 10, pp. 993-1001, 1990.
[http://dx.doi.org/10.1109/34.58871]

[13]   B. Igelnik, Y. H. Pao, S.R. LeClair, and C.Y. Shen, "The ensemble approach to neural-network learning and generalization", *IEEE Trans. Neural Netw.,* vol. 10, no. 1, pp. 19-30, 1999.
[http://dx.doi.org/10.1109/72.737490] [PMID: 18252500]

[14]   N. Ueda, "Optimal linear combination of neural networks for improving classification performance", *IEEE Trans. Pattern Anal. Mach. Intell.,* vol. 22, no. 2, pp. 207-215, 2000.
[http://dx.doi.org/10.1109/34.825759]

[15]   S. Yang, and A. Browne, "Neural network ensembles: combining multiple models for enhanced performance using a multistage approach", *Expert Syst.,* vol. 21, no. 5, pp. 279-288, 2004.
[http://dx.doi.org/10.1111/j.1468-0394.2004.00285.x]

[16]   Xin Yao, and M.M. Islam, "Evolving artificial neural network ensembles", *IEEE Comput. Intell. Mag.,* vol. 3, no. 1, pp. 31-42, 2008.
[http://dx.doi.org/10.1109/MCI.2007.913386]

[17]   P.M. Granitto, P.F. Verdes, H.D. Navone, and H.A. Ceccatto, "Aggregation algorithms for neural network ensemble construction", *VII Proceedings of Brazilian Symposium on Neural Networks,* pp. 178-183, 2003.

[18]   M.M. Islam, Xin Yao, and K. Murase, "A constructive algorithm for training cooperative neural network ensembles", *IEEE Trans. Neural Netw.,* vol. 14, no. 4, pp. 820-834, 2003.
[http://dx.doi.org/10.1109/TNN.2003.813832] [PMID: 18238062]

[19]   A.F. Neto, A.M.P. Canuto, E.F.G. Goldbarg, and M.C. Goldbarg, "Optimization techniques for the selection of members and attributes in ensemble systems", *2011 IEEE Congress of Evolutionary Computation (CEC),* pp. 1912-1919, 2011.
[http://dx.doi.org/10.1109/CEC.2011.5949849]

[20]   M.M. Islam, Xin Yao, S.M. Shahriar Nirjon, M.A. Islam, and K. Murase, "Bagging and boosting negatively correlated neural networks", *IEEE Trans. Syst. Man Cybern. B Cybern.,* vol. 38, no. 3, pp. 771-784, 2008.
[http://dx.doi.org/10.1109/TSMCB.2008.922055] [PMID: 18558541]

[21]   Z-H. Zhou, J-X. Wu, Y. Jiang, and S-F. Chen, "Genetic algorithm based selective neural network

ensemble", *Proceedings of International Joint Conference on Artificial Intelligence,* vol. vol. 2, pp. 797-802, 2001.Seattle

[22]   T. Yu-Bo, Z. Su-Ling, and L. Jing-Yi, "Modeling resonant frequency of microstrip antenna based on neural network ensemble", *Int. J. Numer. Model.,* vol. 24, no. 1, pp. 78-88, 2011.
[http://dx.doi.org/10.1002/jnm.761]

[23]   E.M. Dos Santos, R. Sabourin, and P. Maupin, "Single and multi-objective genetic algorithms for the selection of ensemble of classifiers", *Proceedings of the 2006 IEEE International Joint Conference on Neural Network,* pp. 3070-3077, 2006.

[24]   J. Yang, X. Zeng, S. Zhong, and S. Wu, "Effective neural network ensemble approach for improving generalization performance", *IEEE Trans. Neural Netw. Learn. Syst.,* vol. 24, no. 6, pp. 878-887, 2013.
[http://dx.doi.org/10.1109/TNNLS.2013.2246578] [PMID: 24808470]

[25]   D.S. Weile, and E. Michielssen, "Genetic algorithm optimization applied to electromagnetics: a review", *IEEE Trans. Antenn. Propag.,* vol. 45, no. 3, pp. 343-353, 1997.
[http://dx.doi.org/10.1109/8.558650]

[26]   D.C. Panda, S.S. Pattnaik, B. Khuntia, D.K. Neog, and S. Devi, "Coupling of ANNN with GA for effective optimization of dimensions of rectangular patch antenna on thick substrate microstrip patch antenna on thick substrate", *Proceedings of 6th International Symposium on Antennas, Propagation and EM Theory,* pp. 720-725, 2003.
[http://dx.doi.org/10.1109/ISAPE.2003.1276788]

[27]   O. Ozgun, S. Mutlu, M.I. Aksun, and L. Alatan, "Design of dual-frequency probe-fed microstrip antennas with genetic optimization algorithm", *IEEE Trans. Antenn. Propag.,* vol. 51, no. 8, pp. 1947-1954, 2003.
[http://dx.doi.org/10.1109/TAP.2003.814732]

[28]   D.W. Boeringer, and D.H. Werner, "Particle swarm optimization versus genetic algorithms for phased array synthesis", *IEEE Trans. Antenn. Propag.,* vol. 52, no. 3, pp. 771-779, 2004.
[http://dx.doi.org/10.1109/TAP.2004.825102]

[29]   J. Kennedy, and R. Eberhart, "Particle swarm optimization", *Proceedings of IEEE Conference on Neural Networks,* pp. 1942-1948, 1995.Piscataway
[http://dx.doi.org/10.1109/ICNN.1995.488968]

[30]   J.R. Perez, and J. Basterrechea, "Comparison of different heuristic optimization methods for near-field antenna measurements", *IEEE Trans. Antenn. Propag.,* vol. 55, no. 3, pp. 549-555, 2007.
[http://dx.doi.org/10.1109/TAP.2007.891508]

[31]   J. Robinson, and Y. Rahmat-Samii, "Particle swarm optimization in electromagnetics", *IEEE Trans. Antenn. Propag.,* vol. 52, no. 2, pp. 397-407, 2004.
[http://dx.doi.org/10.1109/TAP.2004.823969]

[32]   G. Ciuprina, D. Ioan, and I. Munteanu, "Use of intelligent-particle swarm optimization in electromagnetics", *IEEE Trans. Magn.,* vol. 38, no. 2, pp. 1037-1040, 2002.
[http://dx.doi.org/10.1109/20.996266]

[33]   Wen-Chung Liu, "Design of a multiband CPW-fed monopole antenna using a particle swarm optimization approach", *IEEE Trans. Antenn. Propag.,* vol. 53, no. 10, pp. 3273-3279, 2005.
[http://dx.doi.org/10.1109/TAP.2005.856339]

[34]   K.M. Passino, "Biomimicry of bacterial foraging for distributed optimization and control", *IEEE Control Syst.,* vol. 22, no. 3, pp. 52-67, 2002.
[http://dx.doi.org/10.1109/MCS.2002.1004010]

[35]   S.V.R.S. Gollapudi, S.S. Pattnaik, O.P. Bajpai, S. Devi, C. Vidya Sagar, P.K. Pradyumna, and K.M. Bakwad, "Bacterial foraging optimization technique to calculate resonant frequency of rectangular microstrip antenna", *Int. J. RF Microw. Comput.-Aided Eng.,* vol. 18, no. 4, pp. 383-388, 2008.

[http://dx.doi.org/10.1002/mmce.20296]

[36]   T. Datta, and I.S. Misra, "A comparative study of optimization techniques in adaptive antenna array processing: The bacteria-foraging algorithm and particle-swarm optimization", *IEEE Antennas Propag. Mag.,* vol. 51, no. 6, pp. 69-81, 2009.
[http://dx.doi.org/10.1109/MAP.2009.5433098]

[37]   L. dos Santos Coelho, C. da Costa Silveira, C.A. Sierakowski, and P. Alotto, "Improved bacterial foraging strategy applied to TEAM workshop benchmark problem", *IEEE Trans. Magn.,* vol. 46, no. 8, pp. 2903-2906, 2010.
[http://dx.doi.org/10.1109/TMAG.2010.2044026]

[38]   N.A. Okaeme, and P. Zanchetta, "Hybrid bacterial foraging optimization strategy for automated experimental control design in electrical drives", *IEEE Trans. Industr. Inform.,* vol. 9, no. 2, pp. 668-678, 2013.
[http://dx.doi.org/10.1109/TII.2012.2225435]

[39]   S. Dasgupta, S. Das, A. Abraham, and A. Biswas, "Adaptive computational chemotaxis in bacterial foraging optimization: An analysis", *IEEE Trans. Evol. Comput.,* vol. 13, no. 4, pp. 919-941, 2009.
[http://dx.doi.org/10.1109/TEVC.2009.2021982]

[40]   J. Robinson, S. Sinton, and Y. Rahmat-Samii, "Particle swarm, genetic algorithm, and their hybrids: optimization of a profiled corrugated horn antenna", *Proceedings of IEEE Antennas and Propagation Society International Symposium,* pp. 314-317, 2002.
[http://dx.doi.org/10.1109/APS.2002.1016311]

[41]   M.F. Pantoja, P. Meincke, and A.R. Bretones, "A hybrid genetic-algorithm space-mapping tool for the optimization of antennas", *IEEE Trans. Antenn. Propag.,* vol. 55, no. 3, pp. 777-781, 2007.
[http://dx.doi.org/10.1109/TAP.2007.891556]

[42]   D.H. Kim, A. Abraham, and J.H. Cho, "A hybrid genetic algorithm and bacterial foraging approach for global optimization", *Inf. Sci.,* vol. 177, no. 18, pp. 3918-3937, 2007.
[http://dx.doi.org/10.1016/j.ins.2007.04.002]

[43]   F. Grimaccia, M. Mussetta, and R.E. Zich, "Genetical swarm optimization: Self-adaptive hybrid evolutionary algorithm for electromagnetics", *IEEE Trans. Antenn. Propag.,* vol. 55, no. 3, pp. 781-785, 2007.
[http://dx.doi.org/10.1109/TAP.2007.891561]

[44]   S.V.R.S. Gollapudi, S.S. Pattnaik, O.P. Bajpai, S. Devi, V. Sagar, P.K. Pradyumna, and K.M. Bakwad, "Hybridized germ swarm optimization technique to calculate resonant frequency of RMA", *Proceedings of International Conference on Recent Advances in Microwave Theory and Applications,* pp. 280-283, 2008.
[http://dx.doi.org/10.1109/AMTA.2008.4763022]

[45]   L.X. Long, and L.R. Jun, "A bacterial foraging global optimization algorithm based on the particle swarm optimization", *Proceedings of IEEE International Conference on Intelligent Computing and Intelligent Systems,* vol. vol. 2, pp. 22-27, 2010.Xiamen

[46]   R.K. Mishra, and A. Patnaik, "Designing rectangular patch antenna using the neurospectral method", *IEEE Trans. Antenn. Propag.,* vol. 51, no. 8, pp. 1914-1921, 2003.
[http://dx.doi.org/10.1109/TAP.2003.814748]

[47]   N. Turker, and F. Gunes, "Artificial neural design of microstrip antennas", *Turk. J. Electr. Eng. Comput. Sci.,* vol. 14, no. 3, pp. 445-453, 2006.

[48]   B. Khuntia, S.S. Pattnaik, D.C. Panda, D.K. Neog, S. Devi, and M. Dutta, "Genetic algorithm with artificial neural networks as its fitness function to design rectangular microstrip antenna on thick substrate", *Microw. Opt. Technol. Lett.,* vol. 44, no. 2, pp. 144-146, 2005.
[http://dx.doi.org/10.1002/mop.20570]

[49]   S. Lebbar, Z. Guennoun, M. Drissi, and F. Riouch, "A compact and broadband microstrip antenna

design using a geometrical-methodology-based artificial neural network", *IEEE Antennas Propag. Mag.,* vol. 48, no. 2, pp. 146-154, 2006.
[http://dx.doi.org/10.1109/MAP.2006.1650854]

[50]   V.S. Chintakindi, S.S. Pattnaik, O.P. Bajpai, and S. Devi, "PSO driven RBFNN for design of equilateral triangular microstrip patch antenna", *Indian J. Radio Space Phys.,* vol. 34, no. 4, pp. 233-237, 2009.

[51]   K. Siakavara, "Artificial neural network based design of a three-layered microstrip circular ring antenna with specified multi-frequency operation", *Neural Comput. Appl.,* vol. 18, no. 1, pp. 57-64, 2009.
[http://dx.doi.org/10.1007/s00521-007-0153-3]

[52]   K.A. Kumar, R. Ashwath, D.S. Kumar, and R. Malmathanraj, "Optimization of multislotted rectangular microstrip patch antenna using ANN and bacterial foraging optimization", *Proceedings of 2010 Asia-Pacific International Symposium on Electromagnetic Compatibility,* pp. 449-452, 2010.
[http://dx.doi.org/10.1109/APEMC.2010.5475810]

[53]   S.S. Gultekin, D. Uzer, and O. Dundar, "Calculation of circular microstrip antenna parameters with a single artificial neural network model", *Proceedings of Progress in Electromagnetics Research Symposium,* pp. 545-548, 2012.

[54]   Z. Wang, S. Fang, Q. Wang, and H. Liu, "An ANN-based synthesis model for the single-feed circularly-polarized square microstrip antenna with truncated corners", *IEEE Trans. Antenn. Propag.,* vol. 60, no. 12, pp. 5989-5992, 2012.
[http://dx.doi.org/10.1109/TAP.2012.2214195]

[55]   T. Bose, and N. Gupta, "Design of an aperture-coupled microstrip antenna using a hybrid neural network", *IET Microw. Antennas Propag.,* vol. 6, no. 4, pp. 470-474, 2012.
[http://dx.doi.org/10.1049/iet-map.2011.0363]

[56]   I. Vilovic, N. Burum, and M. Brailo, "Microstrip antenna design using neural networks optimized by PSO", *Proceedings of International Conference on Applied Electromagnetics and Communications,* pp. 1-4, 2013.
[http://dx.doi.org/10.1109/ICECom.2013.6684759]

[57]   S. Devi, D.C. Panda, S.S. Pattnaik, B. Khuntia, and D.K. Neog, "Initializing artificial neural networks by genetic algorithm to calculate the resonant frequency of single shorting post rectangular patch antenna", *Proceedings of IEEE Antennas and Propagation Society International Symposium,* pp. 144-147, 2003.
[http://dx.doi.org/10.1109/APS.2003.1219810]

[58]   B. Khuntia, S.S. Pattnaik, D.C. Panda, D.K. Neog, S. Devi, and M. Dutta, "A simple and efficient approach to train artificial neural networks using a genetic algorithm to calculate the resonant frequency of an RMA on thick substrate", *Microw. Opt. Technol. Lett.,* vol. 41, no. 4, pp. 313-315, 2004.
[http://dx.doi.org/10.1002/mop.20126]

[59]   S.S. Pattnaik, B. Khuntia, D.C. Panda, D.K. Neog, S. Devi, and M. Dutta, "Application of a genetic algorithm in an artificial neural network to calculate the resonant frequency of a tunable single-shorting-post rectangular-patch antenna", *Int. J. RF Microw. Comput.-Aided Eng.,* vol. 15, no. 1, pp. 140-144, 2005.
[http://dx.doi.org/10.1002/mmce.20060]

[60]   G. Angiulli, and M. Versaci, "Resonant frequency evaluation of microstrip antennas using a neural-fuzzy approach", *IEEE Trans. Magn.,* vol. 39, no. 3, pp. 1333-1336, 2003.
[http://dx.doi.org/10.1109/TMAG.2003.810172]

[61]   K. Guney, and N. Sarikaya, "A hybrid method based on combining artificial neural network and fuzzy inference system for simultaneous computation of resonant frequencies of rectangular, circular, and triangular microstrip antennas", *IEEE Trans. Antenn. Propag.,* vol. 55, no. 3, pp. 659-668, 2007.

[http://dx.doi.org/10.1109/TAP.2007.891566]

[62]   S. P. Gangwar, R. P. S. Gangwar, and B. K. Kanaujia, "Resonant frequency of circular microstrip antenna using artificial neural networks", *Indian J. Radio Space Phys.*, vol. 37, no. 3, pp. 204-208, 2008.

[63]   Y. Tighilt, F. Bouttout, and A. Khellaf, "Modeling and design of printed antennas using neural networks", *Int. J. RF Microw. Comput.-Aided Eng.*, vol. 21, no. 2, pp. 228-233, 2011.
[http://dx.doi.org/10.1002/mmce.20509]

[64]   S. Can, K.Y. Kapusuz, and E. Aydin, "Calculation of resonant frequencies of a shorting pin-loaded ETMA with ANN", *Microw. Opt. Technol. Lett.*, vol. 56, no. 3, pp. 660-663, 2014.
[http://dx.doi.org/10.1002/mop.28157]

[65]   K. Hettak, and G.Y. Delisle, "Low profile cellular radio antenna for ISM applications", *Proceedings of IEEE Antennas and Propagation Society International Symposium*, pp. 443-446, 2004.

[66]   D.K. Neog, S.S. Pattnaik, D.C. Panda, S. Devi, B. Khuntia, and M. Dutta, "Design of a wideband microstrip antenna and the use of artificial neural networks in parameter calculation", *IEEE Antennas Propag. Mag.*, vol. 47, no. 3, pp. 60-65, 2005.
[http://dx.doi.org/10.1109/MAP.2005.1532541]

[67]   K. Guney, and N. Sarikaya, "Comparison of adaptive-network-based fuzzy inference systems for bandwidth calculation of rectangular microstrip antennas", *Expert Syst. Appl.*, vol. 36, no. 2, pp. 3522-3535, 2009.
[http://dx.doi.org/10.1016/j.eswa.2008.02.008]

[68]   D.C. Panda, S.S. Pattnaik, S. Devi, and R.K. Mishra, "Application of FIR-neural network on finite difference time domain technique to calculate input impedance of microstrip patch antenna", *Int. J. RF Microw. Comput.-Aided Eng.*, vol. 20, no. 2, pp. 158-162, 2010.
[http://dx.doi.org/10.1002/mmce.20417]

[69]   I. Vilovic, and N. Burum, "Design and feed position estimation for circular microstrip antenna based on neural network model", *Proceedings of 6th European Conference on Antennas and Propagation (EUCAP)*, pp. 3614-3617, 2012.
[http://dx.doi.org/10.1109/EuCAP.2012.6206281]

[70]   T. Khan, A. De, and M. Uddin, "Prediction of slot-size and inserted air-gap for improving the performance of rectangular microstrip antennas using artificial neural networks", *IEEE Antennas Wirel. Propag. Lett.*, vol. 12, pp. 1367-1371, 2013.
[http://dx.doi.org/10.1109/LAWP.2013.2285381]

[71]   M. Aneesh, J.A. Ansari, A. Singh, Kamakshi, and S.S. Sayeed, "Kamakshi, and S. S. Sayeed, Analysis of microstrip line feed slot loaded patch antenna using artificial neural network,", *Prog. Electromagn. Res. B Pier B*, vol. 58, pp. 35-46, 2014.
[http://dx.doi.org/10.2528/PIERB13111105]

[72]   A. Patnaik, D. Anagnostou, C.G. Christodoulou, and J.C. Lyke, "Modeling frequency reconfigurable antenna array using neural networks", *Microw. Opt. Technol. Lett.*, vol. 44, no. 4, pp. 351-354, 2005.
[http://dx.doi.org/10.1002/mop.20632]

[73]   A. Patnaik, D.E. Anagnostou, R. Mishra, C.G. Christodoulou, and J.C. Lyke, "Applications of neural networks in wireless communications", *IEEE Antennas Propag. Mag.*, vol. 46, no. 3, pp. 130-137, 2004.
[http://dx.doi.org/10.1109/MAP.2004.1374125]

[74]   H.J. Delgado, M.H. Thursby, and F.M. Ham, "A novel neural network for the synthesis of antennas and microwave devices", *IEEE Trans. Neural Netw.*, vol. 16, no. 6, pp. 1590-1600, 2005.
[http://dx.doi.org/10.1109/TNN.2005.852973] [PMID: 16342499]

[75]   F.F. Dubrovka, and D.O. Vasylenko, "Neural-genetic optimization applied to the design of UWB planar antennas", *Proceedings of 4th International Conference on Ultrawideband and Ultrashort*

Impulse Signals, pp. 39-41, 2008.
[http://dx.doi.org/10.1109/UWBUS.2008.4669351]

[76]   M. Mudroch, P. Cerny, P. Hazdra, and M. Mazanek, "UWB dipole antenna optimization with neural network tuned algorithm", *Proceedings of European Conference on Antennas and Propagation,* pp. 1491-1494, 2009.Berlin

[77]   F.F. Dubrovka, and D.O. Vasylenko, "Synthesis of UWB planar antennas by means of natural optimization algorithms", *Radioelectron. Commun. Syst.,* vol. 52, no. 4, pp. 167-178, 2009.
[http://dx.doi.org/10.3103/S0735272709040013]

[78]   L. Lucci, G. Pelosi, and S. Selleri, "An artificial neural network approach for the harmonic design of annular ring dielectric resonator antennas", *Prog. Electromagn. Res. C. Pier C,* vol. 22, pp. 35-45, 2011.
[http://dx.doi.org/10.2528/PIERC11042205]

[79]   K. Sri Rama Krishna, K. Jagadeesh Babu, J. Lakshmi Narayana, L. Pratap Reddy, and G.V. Subrahmanyam, "Artificial neural network approach for analyzing mutual coupling in a rectangular MIMO antenna", *Frontiers of Electrical and Electronic Engineering,* vol. 7, no. 3, pp. 293-298, 2012.
[http://dx.doi.org/10.1007/s11460-012-0203-1]

[80]   F. Gunes, S. Nesil, and S. Demirel, "Design and analysis of Minkowski reflectarray antenna using 3-D CST microwave studio-based neural network model with particle swarm optimization", *Int. J. RF Microw. Comput-Aid. Eng.,* vol. 23, no. 2, pp. 272-284, 2013.

[81]   S.S. Pattnaik, B. Khuntia, D.C. Panda, D.K. Neog, and S. Devi, "Calculation of optimized parameters of rectangular microstrip patch antenna using genetic algorithm", *Microw. Opt. Technol. Lett.,* vol. 37, no. 6, pp. 431-433, 2003.
[http://dx.doi.org/10.1002/mop.10940]

[82]   A.J. Kerkhoff, R.L. Rogers, and H. Ling, "Design and analysis of planar monopole antennas using a genetic algorithm approach", *IEEE Trans. Antenn. Propag.,* vol. 52, no. 10, pp. 2709-2718, 2004.
[http://dx.doi.org/10.1109/TAP.2004.834429]

[83]   C.M. Coleman, E.J. Rothwell, and J.E. Ross, "Investigation of simulated annealing, ant-colony optimization, and genetic algorithms for self-structuring antennas", *IEEE Trans. Antenn. Propag.,* vol. 52, no. 4, pp. 1007-1014, 2004.
[http://dx.doi.org/10.1109/TAP.2004.825658]

[84]   N. Jin, and Y. Rahmat-Samii, "Parallel particle swarm optimization and finite- difference time-domain (PSO/FDTD) algorithm for multiband and wide-band patch antenna designs", *IEEE Trans. Antenn. Propag.,* vol. 53, no. 11, pp. 3459-3468, 2005.
[http://dx.doi.org/10.1109/TAP.2005.858842]

[85]   Z. Lukeš, and Z. Raida, "Multi-objective optimization of wire antennas: genetic algorithms versus particle swarm optimization", *Wuxiandian Gongcheng,* vol. 14, pp. 91-97, 2005.

[86]   H. Choo, R. L. Rogers, and H. Ling, "Design of electrically small wire antennas using a pareto genetic algorithm", *IEEE Trans. Antenn. Propag.,* vol. 53, no. 3, pp. 1038-1046, 2005.
[http://dx.doi.org/10.1109/TAP.2004.842404]

[87]   P. Soontornpipit, C.M. Furse, and Y.C. Chung, "Design of implantable microstrip antenna for communication with medical implants", *IEEE Trans. Microw. Theory Tech.,* vol. 52, no. 8, pp. 1944-1951, 2004.
[http://dx.doi.org/10.1109/TMTT.2004.831976]

[88]   I.S. Misra, R.S. Chakrabarty, and B.B. Mangaraj, "Design, analysis and optimization of v-dipole and its three-element Yagi-Uda array", *Electromagn. waves,* vol. 66, pp. 137-156, 2006.
[http://dx.doi.org/10.2528/PIER06102604]

[89]   N. Telzhensky, and Y. Leviatan, "Novel method of UWB antenna optimization for specified input signal forms by means of genetic algorithm", *IEEE Trans. Antenn. Propag.,* vol. 54, no. 8, pp. 2216-

2225, 2006.
[http://dx.doi.org/10.1109/TAP.2006.879201]

[90]   A.Z. Hood, and E. Topsakal, "Particle swarm optimization for dual-band implantable antennas", *Proceedings of 2007 IEEE Antennas and Propagation Society International Symposium,* pp. 3209-3212, 2007.
[http://dx.doi.org/10.1109/APS.2007.4396219]

[91]   M. Ding, R. Jin, and J. Geng, "Optimal design of ultra wideband antennas using a mixed model of 2-D genetic algorithm and finite-difference time-domain", *Microw. Opt. Technol. Lett.,* vol. 49, no. 12, pp. 3177-3180, 2007.
[http://dx.doi.org/10.1002/mop.22928]

[92]   V.S. Chintakindi, S.S. Pattnaik, O.P. Bajpai, S. Devi, S.V.R.S. Gollapudi, and P.K. Pradyumna, "Parameters calculations of rectangular microstrip patch antenna using Particle Swarm Optimization technique", *Proceedings of 2007 IEEE Applied Electromagnetics Conference (AEMC),* pp. 1-4, 2007.
[http://dx.doi.org/10.1109/AEMC.2007.4638010]

[93]   N. Jin, and Y. Rahmat-Samii, "Advances in particle swarm optimization for antenna designs: real-number, binary, single-objective and multiobjective implementations", *IEEE Trans. Antenn. Propag.,* vol. 55, no. 3, pp. 556-567, 2007.
[http://dx.doi.org/10.1109/TAP.2007.891552]

[94]   M. Rattan, M.S. Patterh, and B.S. Sohi, "Design of a linear array of half wave parallel dipoles using particle swarm optimization", *Prog. Electromagn. Res. M Pier M,* vol. 2, pp. 131-139, 2008.
[http://dx.doi.org/10.2528/PIERM08040702]

[95]   T. Datta, I.S. Misra, B.B. Mangaraj, and S. Imtiaj, "Improved adaptive bacteria foraging algorithm in optimization of antenna array for faster convergence", *Prog. Electromagn. Res. C. Pier C,* vol. 1, pp. 143-157, 2008.
[http://dx.doi.org/10.2528/PIERC08011705]

[96]   M. Shihab, Y. Najjar, N. Dib, and M. Khodier, "Design of non-uniform circular antenna arrays using particle swarm optimization", *J. Electr. Eng.,* vol. 59, no. 4, pp. 216-220, 2008.

[97]   Y. Cengiz, and H. Tokat, "Linear antenna array design with use of genetic, memetic and Tabu search optimization algorithms", *Prog. Electromagn. Res. C. Pier C,* vol. 1, pp. 63-72, 2008.
[http://dx.doi.org/10.2528/PIERC08010205]

[98]   V.S. Chintakindi, S.S. Pattnaik, O.P. Bajpai, S. Devi, P.K. Patra, and K.M. Bakwad, "Resonant frequency of equilateral triangular microstrip patch antenna using particle swarm optimization technique", *Proceedings of 2008 International Conference on Recent Advances in Microwave Theory and Applications,* pp. 20-22, 2008.
[http://dx.doi.org/10.1109/AMTA.2008.4763004]

[99]   N. Jin, and Y. Rahmat-Samii, "Particle swarm optimization for antenna designs in engineering electromagnetics", *J. Artif. Evol. Appl.,* vol. 2008, pp. 1-10, 2008.
[http://dx.doi.org/10.1155/2008/728929]

[100] T. Karacolak, A.Z. Hood, and E. Topsakal, "Design of a dual-band implantable antenna and development of skin mimicking gels for continuous glucose monitoring", *IEEE Trans. Microw. Theory Tech.,* vol. 56, no. 4, pp. 1001-1008, 2008.
[http://dx.doi.org/10.1109/TMTT.2008.919373]

[101] M.A. Panduro, C.A. Brizuela, L.I. Balderas, and D.A. Acosta, "A comparison of genetic algorithms, particle swarm optimization and the differential evolution method for the design of scannable circular antenna arrays", *Prog. Electromagn. Res. B Pier B,* vol. 13, pp. 171-186, 2009.
[http://dx.doi.org/10.2528/PIERB09011308]

[102] S.V.R.S. Gollapudi, S.S. Pattnaik, O.P. Bajapai, S. Devi, K.M. Bakwad, and P.K. Pradyumna, "Intelligent bacterial foraging optimization technique to calculate resonant frequency of RMA", *Int. J. Microw. Opt. Technol.,* vol. 4, no. 2, pp. 67-75, 2009.

[103] H. Wu, J. Geng, R. Jin, J. Qiu, W. Liu, J. Chen, and S. Liu, "An improved comprehensive learning particle swarm optimization and its application to the semiautomatic design of antennas", *IEEE Trans. Antenn. Propag.*, vol. 57, no. 10, pp. 3018-3028, 2009.
[http://dx.doi.org/10.1109/TAP.2009.2028608]

[104] K.R. Mahmoud, "Design optimization of a bow-tie antenna for 2.45 GHz RFID readers using a hybrid BSO-NM algorithm", *Electromagn. waves*, vol. 100, pp. 105-117, 2010.
[http://dx.doi.org/10.2528/PIER09102903]

[105] Lin-Yu Tseng, and Tuan-Yung Han, "An evolutionary design method using genetic local search algorithm to obtain broad/dual-band characteristics for circular polarization slot antennas", *IEEE Trans. Antenn. Propag.*, vol. 58, no. 5, pp. 1449-1456, 2010.
[http://dx.doi.org/10.1109/TAP.2010.2044312]

[106] N. Jin, and Y. Rahmat-Samii, "Hybrid real-binary particle swarm optimization (HPSO) in engineering electromagnetics", *IEEE Trans. Antenn. Propag.*, vol. 58, no. 12, pp. 3786-3794, 2010.
[http://dx.doi.org/10.1109/TAP.2010.2078477]

[107] S. Sharma, and B.K. Kanaujia, "Optimization of resonant frequency of circular microstrip antenna with and without air gaps using bacterial foraging optimization technique", *Proceedings of 2011 International Conference on Computational Intelligence and Communication Networks*, pp. 574-577, 2011.
[http://dx.doi.org/10.1109/CICN.2011.123]

[108] S.H. Yeung, W.S. Chan, K.T. Ng, and K.F. Man, "Computational optimization algorithms for antennas and RF/microwave circuit designs: An overview", *IEEE Trans. Industr. Inform.*, vol. 8, no. 2, pp. 216-227, 2012.
[http://dx.doi.org/10.1109/TII.2012.2186821]

[109] V. Anjitha, and S. Kumar, "Optimal design of Zig-Zag antenna using nonlinear segment length and pitch angle", *Procedia Technol.*, vol. 6, pp. 799-805, 2012.
[http://dx.doi.org/10.1016/j.protcy.2012.10.097]

[110] C.R.M. Silva, H.W.C. Lins, S.R. Martins, E.L.F. Barreto, and A.G. d'Assunção, "A multiobjective optimization of a UWB antenna using a self organizing genetic algorithm", *Microw. Opt. Technol. Lett.*, vol. 54, no. 8, pp. 1824-1828, 2012.
[http://dx.doi.org/10.1002/mop.26945]

[111] Y. Rahmat-Samii, J.M. Kovitz, and H. Rajagopalan, "Nature-inspired optimization techniques in communication antenna designs", *Proc. IEEE*, vol. 100, no. 7, pp. 2132-2144, 2012.
[http://dx.doi.org/10.1109/JPROC.2012.2188489]

[112] A.A. Minasian, and T.S. Bird, "Particle swarm optimization of microstrip antennas for wireless communication systems", *IEEE Trans. Antenn. Propag.*, vol. 61, no. 12, pp. 6214-6217, 2013.
[http://dx.doi.org/10.1109/TAP.2013.2281517]

[113] Y.L. Li, W. Shao, L. You, and B.Z. Wang, "An improved PSO algorithm and its application to UWB antenna design", *IEEE Antennas Wirel. Propag. Lett.*, vol. 12, pp. 1236-1239, 2013.
[http://dx.doi.org/10.1109/LAWP.2013.2283375]

[114] C.R.M. Silva, and S.R. Martins, "An adaptive evolutionary algorithm for UWB microstrip antennas optimization using a machine learning technique", *Microw. Opt. Technol. Lett.*, vol. 55, no. 8, pp. 1864-1868, 2013.
[http://dx.doi.org/10.1002/mop.27692]

[115] M. Cismasu, M. Gustafsson, G. Ciuprina, D. Ioan, and I. Munteanu, "An improved PSO algorithm and its application to UWB antenna design", *IEEE Trans. Antenn. Propag.*, vol. 62, no. 3, pp. 1037-1040, 2002.

[116] A. Deb, J.S. Roy, and B. Gupta, "Performance comparison of differential evolution, particle swarm optimization and genetic algorithm in the design of circularly polarized microstrip antennas", *IEEE*

Trans. Antenn. Propag., vol. 62, no. 8, pp. 3920-3928, 2014.
[http://dx.doi.org/10.1109/TAP.2014.2322880]

[117] L.H. Manh, F. Grimaccia, M. Mussetta, and R.E. Zich, "Optimization of a dual ring antenna by means of artificial neural network", *Prog. Electromagn. Res. B Pier B,* vol. 58, pp. 59-69, 2014.
[http://dx.doi.org/10.2528/PIERB13112806]

[118] N.R.S.V.P. Koppisetti, S. Mallick, and V.K. Bhargava, "Design of adaptive antenna systems for LTE using genetic algorithm and particle swarm optimization", *Proceedings of 2015 IEEE 28th Canadian Conference on Electrical and Computer Engineering (CCECE),* pp. 1054-1059, 2015.
[http://dx.doi.org/10.1109/CCECE.2015.7129420]

[119] D.H. Werner, P.L. Werner, and K.H. Church, "Genetically engineered multiband fractal antennas", *Electron. Lett.,* vol. 37, no. 19, pp. 1150-1151, 2001.
[http://dx.doi.org/10.1049/el:20010802]

[120] D.H. Werner, P.L. Werner, J.W. Culver, S.D. Eason, and R. Libonati, "Load sensitivity analysis for genetically engineered miniature multiband fractal dipole antennas", *Proceedings of IEEE Antennas and Propagation Society International Symposium,* pp. 86-89, 2002.
[http://dx.doi.org/10.1109/APS.2002.1016932]

[121] M. Fernandez Pantoja, F. Garcia Ruiz, A. Rubio Bretones, R. Gomez Martin, J.M. Gonzalez-Arbesu, J. Romeu, and J.M. Rius, "GA design of wire pre-fractal antennas and comparison with other Euclidean geometries", *IEEE Antennas Wirel. Propag. Lett.,* vol. 2, pp. 238-241, 2003.
[http://dx.doi.org/10.1109/LAWP.2003.819694]

[122] D.H. Werner, and P.L. Werner, "The design optimization of miniature low profile antennas placed in close proximity to high-impedance surfaces", *Proceedings of IEEE Antennas and Propagation Society International Symposium,* pp. 157-160, 2003.
[http://dx.doi.org/10.1109/APS.2003.1217424]

[123] P. Dehkhoda, and A. Tavakoli, "A crown square microstrip fractal antenna", *Proceedings of IEEE Antennas and Propagation Society Symposium,* pp. 2396-2399, 2004.
[http://dx.doi.org/10.1109/APS.2004.1331855]

[124] T.G. Spence, and D.H. Werner, "Genetically optimized fractile microstrip patch antennas", *Proceedings of IEEE Antennas and Propagation Society Symposium,* pp. 4424-4427, 2004.
[http://dx.doi.org/10.1109/APS.2004.1330333]

[125] R. Azaro, G. Boato, M. Donelli, G. Franceschini, A. Martini, and A. Massa, "Design of miniaturised ISM-band fractal antenna", *Electron. Lett.,* vol. 41, no. 14, p. 785, 2005.
[http://dx.doi.org/10.1049/el:20050774]

[126] R. Azaro, G. Boato, M. Donelli, A. Massa, and E. Zeni, "Design of a prefractal monopolar antenna for 3.4-3.6 GHz WI-max band portable devices", *IEEE Antennas Wirel. Propag. Lett.,* vol. 5, pp. 116-119, 2006.
[http://dx.doi.org/10.1109/LAWP.2006.872427]

[127] M.F. Pantoja, F.G. Ruiz, A.R. Bretones, S.G. Garcia, R.G. Martín, J.M.G. Arbesu, J. Romeu, J.M. Rius, P.L. Werner, and D.H. Werner, "GA design of small thin-wire antennas: Comparison with Sierpinsky-type prefractal antennas", *IEEE Trans. Antenn. Propag.,* vol. 54, no. 6, pp. 1879-1882, 2006.
[http://dx.doi.org/10.1109/TAP.2006.875931]

[128] M. Polpasee, N. Homsup, and P. Virunha, "Optimized directivity pattern for arrays by using genetic algorithms based on planar fractal arrays", *Proceedings of IEEE International Symposium on Communications and Information Technologies,* pp. 28-31, 2006.Bangkok

[129] D. Franceschini, R. Azaro, L. Manica, and A. Massa, "A miniaturization process of an antenna with pre-fractal geometry by means of a particle swarm optimization", *Proceedings of 2006 IEEE Antennas and Propagation Society International Symposium,* pp. 3539-3542, 2006.
[http://dx.doi.org/10.1109/APS.2006.1711382]

[130] R. Ghatak, R.K. Mishra, and D.R. Poddar, "Optimization of a Sierpinski carpet pre-fractal planar monopole antenna using real coded genetic algorithm for wideband application", *Proceedings of 2007 IEEE Applied Electromagnetics Conference (AEMC)*, pp. 1-4, 2007.
[http://dx.doi.org/10.1109/AEMC.2007.4638018]

[131] R. Ghatak, D.R. Poddar, and R.K. Mishra, "Design of Sierpinski gasket fractal microstrip antenna using real coded genetic algorithm", *IET Microw. Antennas Propag.*, vol. 3, no. 7, pp. 1133-1140, 2009.
[http://dx.doi.org/10.1049/iet-map.2008.0257]

[132] R. Azaro, E. Zeni, P. Rocca, and A. Massa, "Design of non-harmonic multi-band pre-fractal antennas", *Proceedings of 2007 IEEE Antennas and Propagation Society International Symposium*, pp. 1613-1616, 2007.
[http://dx.doi.org/10.1109/APS.2007.4395819]

[133] R. Azaro, E. Zeni, P. Rocca, and A. Massa, "Innovative design of a planar fractal-shaped GPS/GSM/Wi-Fi antenna", *Microw. Opt. Technol. Lett.*, vol. 50, no. 3, pp. 825-829, 2008.
[http://dx.doi.org/10.1002/mop.23208]

[134] P. Rocca, R. Azaro, M. Benedetti, F. Viani, E. Zeni, and A. Massa, "Multi-band patch antenna tuning by a fractal-shaped erosion process", *Proceedings of 2008 IEEE Antennas and Propagation Society International Symposium*, pp. 1-4, 2008.
[http://dx.doi.org/10.1109/APS.2008.4619445]

[135] B.M. Vidal, and Z. Raida, "Synthesizing Sierpinski Antenna by genetic algorithm and swarm optimization", *Wuxiandian Gongcheng*, vol. 17, pp. 25-29, 2008.

[136] R. Azaro, L. Debiasi, E. Zeni, M. Benedetti, P. Rocca, and A. Massa, "A hybrid prefractal three-band antenna for multistandard mobile wireless applications", *IEEE Antennas Wirel. Propag. Lett.*, vol. 8, pp. 905-908, 2009.
[http://dx.doi.org/10.1109/LAWP.2009.2028627]

[137] L. Lizzi, F. Viani, E. Zeni, and A. Massa, "A DVBH/GSM/UMTS planar antenna for multimode wireless devices", *IEEE Antennas Wirel. Propag. Lett.*, vol. 8, pp. 568-571, 2009.
[http://dx.doi.org/10.1109/LAWP.2009.2022962]

[138] P. Hazdra, M. Capek, and J. Kracek, "Optimization tool for fractal patches based on the IFS algorithm", *Proceedings of European Conference on Antennas and Propagation*, pp. 1837-1839, 2009.Berlin

[139] N.D. Hieu, D.N. Chien, and N.K. Kiem, "Flexible PSO-based optimization of millimeter-wave triple-band antennas by the use of fractal configuration", *Proceedings of the 2010 International Conference on Advanced Technologies for Communications*, pp. 336-340, 2010.
[http://dx.doi.org/10.1109/ATC.2010.5672676]

[140] E.C.D. Oliveira, A.G. Assuncao, and C.R.M.D. Silva, "Optimization of the input impedance of Koch triangular quasi-fractal antennas using genetic algorithms", *Proceedings of IEEE Conference on Electromagnetic Field Computation*, p. 1, 2010.Chicago
[http://dx.doi.org/10.1109/CEFC.2010.5481791]

[141] M. Capek, P. Hazdra, P. Hamouz, and M. Mazanek, "Software tools for efficient generation, modelling and optimisation of fractal radiating structures", *IET Microw. Antennas Propag.*, vol. 5, no. 8, pp. 1002-1007, 2011.
[http://dx.doi.org/10.1049/iet-map.2010.0269]

[142] M. Capek, and P. Hazdra, "Design of IFS patch antenna using particle swarm optimization", *Proceedings of the Fourth European Conference on Antennas and Propagation*, pp. 1-5, 2010.

[143] Z. Adelpour, F. Mohajeri, and M. Sadeghi, "Dual-frequency microstrip patch antenna with modified Koch fractal geometry based on genetic algorithm", In: *Proceedings of 2010 Loughborough Antennas & Propagation Conference*, 2010, pp. 401-404.

[http://dx.doi.org/10.1109/LAPC.2010.5666299]

[144] Anuradha, A. Patnaik, and S.N. Sinha, "Design of custom-made fractal multi-band antennas using ANN-PSO", *IEEE Antennas Propag. Mag.,* vol. 53, no. 4, pp. 94-101, 2011.
[http://dx.doi.org/10.1109/MAP.2011.6097296]

[145] R.O. Ouedraogo, E.J. Rothwell, A.R. Diaz, K. Fuchi, and A. Temme, "Miniaturization of patch antennas using a metamaterial-inspired technique", *IEEE Trans. Antenn. Propag.,* vol. 60, no. 5, pp. 2175-2182, 2012.
[http://dx.doi.org/10.1109/TAP.2012.2189699]

[146] L. Lizzi, and A. Massa, "Dual-band printed fractal monopole antenna for LTE applications", *IEEE Antennas Wirel. Propag. Lett.,* vol. 10, pp. 760-763, 2011.
[http://dx.doi.org/10.1109/LAWP.2011.2163051]

[147] J.B. Krishna, P. Parvathi, and N. Latha, "Adaptive neuro fuzzy inference system for the optimization of behavior of fractal antenna", *Proceedings of 2012 International Conference on Communications, Devices and Intelligent Systems (CODIS),* pp. 242-245, 2012.
[http://dx.doi.org/10.1109/CODIS.2012.6422183]

[148] W.C. Weng, and C.L. Hung, "An H-fractal antenna for multiband applications", *IEEE Antennas Wirel. Propag. Lett.,* vol. 13, pp. 1705-1708, 2014.
[http://dx.doi.org/10.1109/LAWP.2014.2351618]

[149] R. Ghatak, A. Karmakar, and D.R. Poddar, "Evolutionary optimization of Haferman carpet fractal patterned antenna array", *Int. J. RF Microw. Comput.-Aided Eng.,* vol. 25, no. 8, pp. 719-729, 2015.
[http://dx.doi.org/10.1002/mmce.20911]

CHAPTER 3

Fractal Antennas

Abstract: This chapter discusses fractal geometry concepts and fractal antennas. Selected fractal antennas and their features are described, and all the designed fractal antennas are introduced in this chapter. The important features like miniaturization & multiband operation of the designed fractal antennas are highlighted, and their applications are also discussed.

Keywords: Crown fractal antenna, Fractal antenna, Miniaturized antenna, Sierpinski gasket.

INTRODUCTION

The term 'fractal' is used to represent a class of geometry with unique properties. This term was originally employed by Mandelbrot to describe recursively generated self-similar geometric shapes [1]. The shape of fractal structures is similar at different scales, *i.e.*, the subparts of the overall geometry are similar to the overall global shape. Due to this property, the fractal shapes are called self-similar shapes [2]. Another important property of fractal geometries is the space-filling property, which means the possibility to enclose an infinitely long curve in a finite area [3] *e.g.*, a surface with a very large perimeter can be enclosed in fractal loops.

Another important property of fractal shapes is the fractional dimensions, *i.e.*, the dimensions of the fractal geometry are not whole numbers but fractional numbers [1]. The self-similarity property of fractals states that the fractional dimension (*Dim*) of fractal geometry is calculated by the equation given below [4]:

$$Dim = \frac{\log(M)}{\log(s_d)} \qquad (3.1)$$

where M represents the number of similar copies of fractal geometry and s_d is the inverse of the scaling down ratio of fractal geometry, which is calculated as $1/s_d$ [5, 6].

Many fractal shapes are generated by applying repetitive procedures, such as multiple reduction copy machine algorithm [7] or by using an IFS [8]. In these repetitive algorithms, a starting base shape named 'Initiator' is selected, and another shape named 'Generator' is copied a number of times at various locations, scaling ratios and orientations, to achieve the end geometry [7]. Ideally, fractal geometries are designed by iterating an infinite number of times, however, practically, few starting geometries (also called pre-fractals) are considered [9]. The repetitive designing technique is shown in Figs. (**3.1** and **3.2**) for two different geometries. Fig. (**3.1**) shows the repetitive procedure for the implementation of the Sierpinski gasket geometry. The initiator triangle is replicated at different scales and shifted to form the various iterations. Each triangle is replaced with the three small triangles arranged as shown in Fig. (**3.1**). The repetitive-designing procedure for the Minkowski Island fractal shown in Fig. (**3.2**) involves the replacement of each straight line of the structure by the generator shape.

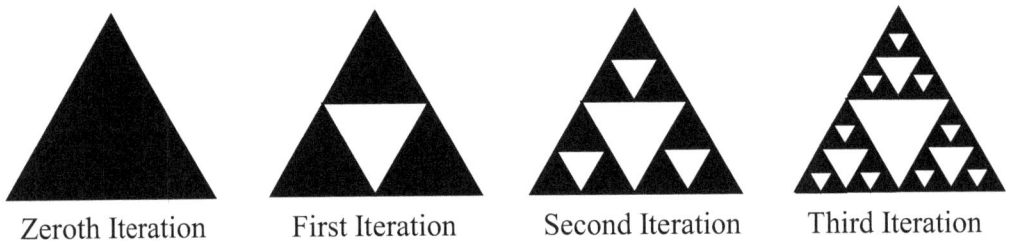

| Zeroth Iteration | First Iteration | Second Iteration | Third Iteration |

Fig. (3.1). Development of Sierpinski Gasket Geometry (Reprinted from the Springer Nature: Neural Computing and Applications, Performance Comparison of Bio-Inspired Optimization Algorithms for Sierpinski Gasket Fractal Antenna Design, Dhaliwal, B.S. and Pattnaik, S.S. © 2016).

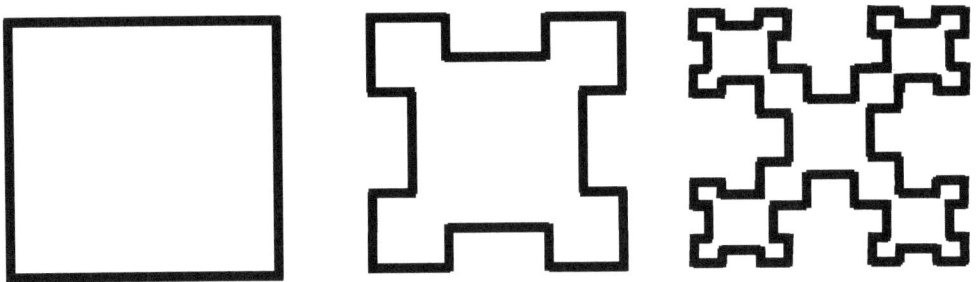

Fig. (3.2). Development of Minkowski Island Fractal Geometry [9].

Fractal geometries have been used in antenna structures to extend antenna design concepts beyond Euclidean geometry [6]. The fractal antennas are antennas in which the fractal geometry concepts are used to design the radiating shapes [10]. The properties of fractal geometries result in a number of advantages in antenna design. The space-filling property of fractal shapes leads to the fitting of very long electrical lengths into compact physical spaces. This results in the miniaturization of antennas [9]. The self-similar property of fractals means that various segments of the structure are like the other parts but at different scales. This leads to the multiband behaviour of antennas [11]. The fractional dimensions of the fractals are considered an important mathematical property [12]. In self-similar shapes, the variation of the scale ratio results in changed fractal dimensions of the geometry [5]. There exists a direct relationship between antenna characteristics and variation of dimensions [6]. This property is used for tuning fractal antennas for desired frequencies. The other advantages of fractal antennas include enhanced bandwidth, improved gain and directivity, and better efficiency [13, 14].

SELECTED TYPICAL FRACTAL ANTENNAS AND THEIR FEATURES

Some of the popular fractal antennas are introduced in this section. The electromagnetic behaviour and important features are also described to highlight the potential/advantages of the fractal antennas.

Sierpinski Gasket Monopole Fractal (SGMF) Antenna

The most popular fractal antenna is an antenna based on a fractal shape named as Sierpinski triangle. This fractal shape, also called the Sierpinski gasket, was proposed in 1915 by a Polish mathematician named Waclaw Sierpinski. This fractal shape is used to design a monopole antenna by Puente-Baliarda *et al.* [7], and the fractal antenna is named as the SGMF antenna. The first four iterations of the Sierpinski gasket are shown in Fig. (**3.1**). The starting shape, *i.e.*, the zeroth iteration, is an equilateral triangle from which the first iteration gasket is constructed by subtracting the central inverted triangle. After the subtraction, three equal-sized triangles remain on the structure, each one being half of the size of the original. So, the first iteration shape is a self-similar structure and each one of its three main parts has exactly the same shape as the whole object but is reduced by a factor of two. Each triangle of the first iteration geometry is replaced by three small size rectangles to obtain the second iteration and similarly, this iterative procedure is carried out further to obtain the next iterations [7].

The SGMF monopole antenna is constructed by printing the Sierpinski gasket on a substrate and then mounted over a conducting surface perpendicular to the plane of the printed substrate, which acts as a ground plane. This arrangement is similar to a monopole feeding structure. The geometry is excited by a coaxial signal source from the reverse side of the ground plane.

The SGMF antenna is a self-similar structure and this property is also observed in its S_{11} results and the radiation patterns [7]. This antenna has a multiband performance, and the number of bands depends on the number of iterations *n*. The zeroth iteration has a single fundamental resonant frequency. The first iteration has two resonant frequencies, the second iteration has three resonant frequencies and so on. Also, the spacing between the bands is equal to the scale factor existing among similar structures on the fractal shape, *i.e.*, a log-periodic factor of two.

Sierpinski Carpet Fractal Antenna

The fractal antennas are also designed using the rectangular Sierpinski shape, also known as the Sierpinski carpet. The zeroth iteration, *i.e.*, the base shape, is a square. The base shape is divided into nine equal squares, from which the central square is removed to obtain the first iteration. In the first iteration geometry, eight squares are left, and every square is again partitioned into nine equal squares, and the central square is erased to design the second iteration. This procedure is repeated to obtain next iterations. The Sierpinski carpet geometry is printed on a grounded substrate and fed with a transmission line feeding to implement a fractal antenna by Wong *et al.* [15]. The simulation and experimental results depict that this antenna has a multiband performance. The monopole configuration of this geometry results in a multiband and wideband performance [4].

Koch Curve Antenna

The Koch monopole antenna is also an effective example to demonstrate that fractals can improve some features of common Euclidean shapes. The zeroth iteration, *i.e.*, the first element is a straight segment and the first iteration is obtained by applying the four similarity transformations known as Affine transformation to zeroth iteration [3]. The next iterations are obtained by applying the same Affine transformation iteratively. The length of the curve increases by a factor of 4/3 in each iteration, but maintains exactly the same height [16].

A Koch curve monopole antenna of height 6 cm is analyzed by Puente *et al.* [16] and Baliarda *et al.* [3]. It is found that the Q of this antenna reaches the fundamental limit for small antennas when the number of iterations is increased. However, the measurement of the input return loss of the Koch curve monopole over a wide frequency range suggests that this fractal antenna has harmonic behavior rather than multiband behavior. This antenna is very useful in applications where the reduction of the antenna size is an ultimate goal.

Hexagonal Fractal Antenna

Another example of fractal geometry is a hexagonal fractal antenna. The base shape is a hexagon, and the first iteration shape is designed by grouping six hexagons of size one-third of the original size. Similarly, further iterations are obtained by applying this procedure. A dipole antenna is implemented using these hexagonal fractal shapes, and the corner feeding is used for excitation by Tang and Wahid [17]. The reflection coefficients of the hexagonal fractal antenna show that it resonates in a number of bands like SGMF antenna. However, due to the self-similarity factor of 1/3, the resonant frequencies repeat by a factor of three. The hexagonal fractal antenna is useful in wireless applications where a broader frequency separation is desired.

Crown Square Fractal Antenna

This fractal antenna, named the crown square fractal antenna, is another example of a self-similar shape based on a square shape. To obtain the kth iteration geometry, the size of the $(k-1)^{th}$ iteration geometry is reduced by $1/2^k$, and then this is merged to the $(k-1)^{th}$ iteration geometry.

The results of the crown square fractal antenna show that it has a circular polarization and multiband performance [18]. When compared to a simple square antenna, a size reduction of 16.84% is observed in this crown square fractal antenna.

Rectangular Sierpinski Carpet Based Fractal Antenna

Rectangular Sierpinski carpet geometry-based dual-band and triple-band fractal microstrip with enhanced gain are proposed by Nhlengethwa and Kumar [19]. The gain of the antennas has been enhanced using the defected ground structure and a reflector plane. The experimental verification of the results is also reported.

Sierpinski-Koch Hybrid Fractal Antenna

A miniaturized fractal patch antenna was designed by Chen *et al.* [20] by combining two fractal geometries: the Koch curve and the Sierpinski carpet. The Koch curves are used to etch the edges of the patch and the Sierpinski carpet shape is used for the inner patch. Using this hybrid geometry, a size miniaturization of 77.1% is achieved, along with an improvement in the operating frequency bandwidth. The microstrip line feed in the form of an impedance transformer is used to excite the antenna.

Other Mathematical Fractal Geometries-Based Antenna

The applications of popular mathematical structures such as Mandelbrot, Hilbert, Cantor, Minkowski, Peano, *etc.*, as antennas have been discussed in detail [21], along with the introduction to the mathematicians who developed those shapes. The suitability of these fractal geometries to obtain compact, multiband, antenna arrays, and frequency-selective surfaces has also been described.

FRACTAL ANTENNAS DEVELOPED IN THE PRESENT RESEARCH WORK

A number of fractal antennas have been proposed in the last decade. The fractal antennas have been designed for improved performances such as multiband operation, enhanced gain, miniaturization, *etc*. The main motive of the presented research work is to develop fractal antennas with simple structures having size reduction capabilities. This section introduces the fractal antennas developed in this research work. The detailed analysis and design are discussed in subsequent chapters.

Miniaturized Crown Rectangular Fractal (CRF) Antenna

The first fractal antenna designed in this proposed research work is based on a rectangular shape and is named the CRF antenna because of its similarity to the antenna proposed by Dehkhoda and Tavakoli [18]. The realization of this fractal antenna is shown in Fig. (**3.3**).

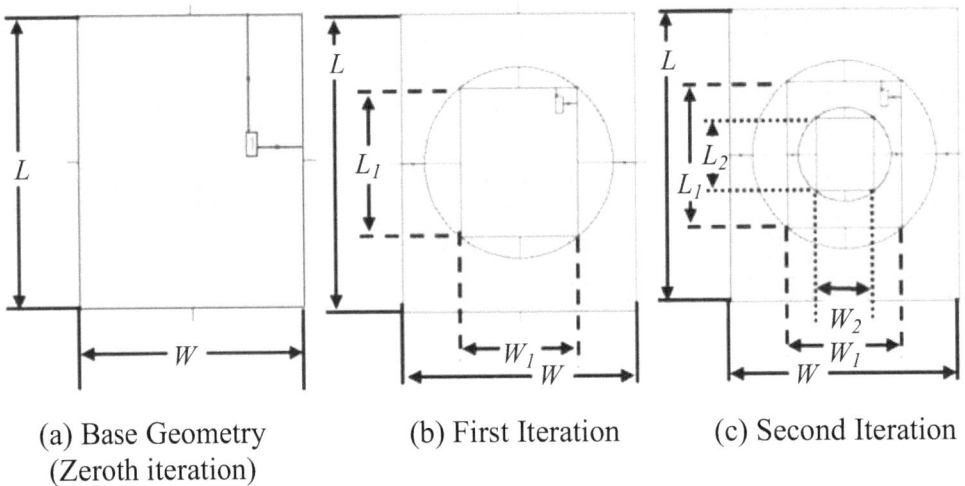

(a) Base Geometry
(Zeroth iteration)

(b) First Iteration

(c) Second Iteration

Fig. (3.3). Proposed CRF Antenna.

The base geometry, *i.e.*, zeroth iteration, is a rectangle, as shown in Fig. (**3.3(a)**). The first iteration geometry, shown in Fig. (**3.3(b)**), is obtained by cutting an ellipse from the base shape and then inserting a rectangle such that the corners of the inserted rectangle touch the boundary of the elliptical slot. The same procedure is repeated for the inner rectangle of the first iteration geometry to obtain the second iteration geometry which is shown in Fig. (**3.3(c)**). Similarly, further iterations can be obtained.

The base geometry shown in Fig. (**3.3(a)**) has two design variables: the length L and the width W of the antenna; Fig. (**3.3(b)**) depicts that the first iteration shape has four variables: the lengths L & L_1 and the widths W & W_1; and the second iteration geometry has six variables: the lengths L, L_1 & L_2 and the widths W, W_1 & W_2 as shown in Fig. (**3.3(c)**). In addition, the resonant frequency also depends on the substrate parameters, *i.e.*, the height of substrate h, dielectric constant ε_r of the substrate, and also on copper patch thickness t. Therefore, the resonant frequency of the second iteration of the proposed fractal antenna depends on eight parameters assuming $t \ll \lambda$ (the wavelength). Consequently, its design for a user-defined frequency requires the optimal values of all these parameters. So, the design of the proposed antenna is a multivariable problem.

However, to reduce the number of design variables and hence, to simplify the design procedure, size of the rectangle inserted in first iteration is selected as 50% of the size of base rectangle and size of the rectangle inserted in the second iteration is 50% of the size of the first iteration rectangle. These assumptions make the number of design variables reduce to four, *i.e.*, the L and W of the base rectangle and the two substrate parameters. Further, in the presented design, the RT-Duroid substrate with $h = 3.175$ mm and $\varepsilon_r = 2.2$ are used. Therefore, the resonant frequency finally depends only on two parameters: the L and W of the base rectangle. The various dimensions of the first iteration and second iteration geometry are calculated from the base rectangle dimensions using the above assumptions. The feed location of the antenna is found by the trial-and-error approach; however, it can also be taken as another variable during antenna design.

In the presented work, the dimensions of the starting shape, *i.e.*, the zeroth iteration are taken as $L = 30$ mm and $W = 37$ mm, respectively. The dimensions of the first and second-iteration geometries are calculated using the above assumptions and are given as follows:

Base geometry (zeroth iteration) dimensions:

- Length of base rectangle $(L) = 30$ mm
- Width of base rectangle $(W) = 37$ mm

First iteration dimensions:

- Length of rectangle to be inserted $(L_1) = 15$ mm
- Width of rectangle to be inserted $(W_1) = 18.5$ mm
- Primary axis radius of ellipse to be cut $= 12$ mm
- Secondary axis radius of ellipse to be cut $= 11.8495$ mm

Second iteration dimensions:

- Length of rectangle to be inserted (L_2) = 7.5 mm
- Width of rectangle to be inserted (W_2) = 9.25 mm
- Primary axis radius of ellipse to be cut = 6 mm
- Secondary axis radius of ellipse to be cut = 5.9247 mm

After a number of trials, the feeding point of base geometry is selected as (8, 2), and that of first-iteration and second-iteration geometries is selected as (5, 7) with respect to the center of all geometries at (0, 0).

The fractal antennas described in Fig. (**3.3**) are simulated using the IE3D software. The S_{11} plots shown in Fig. (**3.4**) depict that the resonance frequency of the base geometry is 3.12 GHz, and that of the first and second iterations is 2.35 GHz and 2.32 GHz, respectively. So, the iterated shapes result in the shifting of resonant frequency values towards the lower end of the frequency scale.

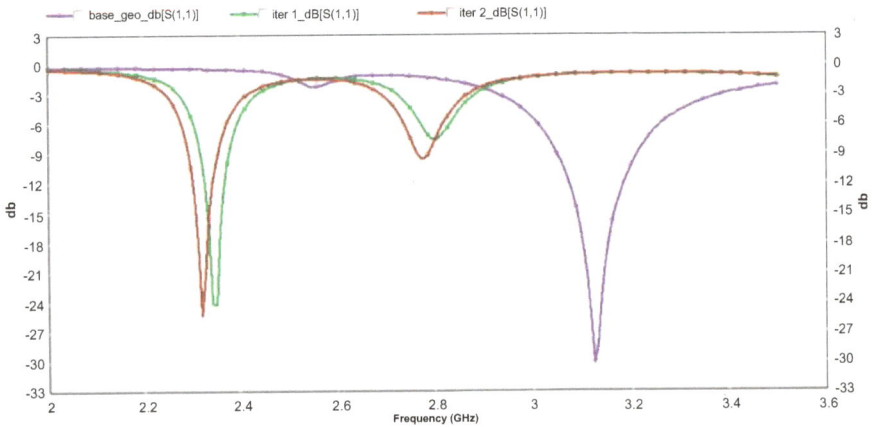

Fig. (3.4). S_{11} Results of CRF Antenna.

The simulated elevation and azimuthal radiation patterns of the antennas shown in Fig. (**3.3**) reveal that the values of peak gain for the base, first and second iteration geometries are 6.97 dBi, 5.59 dBi, and 5.54 dBi, respectively, which are sufficient for low power application in this frequency range. The patterns also show that the CRF antenna has omni-directional radiation patterns. Among the proposed three geometries, the base geometry has a more metallic area, followed by first iteration geometry and then the second iteration geometry. The values of gain also follow this order as the gain directly depends upon the metallic area of the antenna.

The L and W of simple rectangular microstrip antenna resonating at a frequency f_r are calculated using the standard equations [22] described below.

$$W = \frac{1}{2f_r\sqrt{\mu_0\varepsilon_0}}\sqrt{\frac{2}{\varepsilon_e + 1}} = \frac{c}{2f_r}\sqrt{\frac{2}{\varepsilon_r + 1}} \qquad (3.2)$$

$$\varepsilon_{reff} = \frac{\varepsilon_r + 1}{2} + \frac{\varepsilon_r - 1}{2}\left[1 + 10\frac{h}{W}\right]^{-1/2} \qquad (3.3)$$

$$\Delta L = 0.412h\frac{(\varepsilon_{reff} + 0.3)\left(\frac{W}{h} + 0.264\right)}{(\varepsilon_{reff} - 0.258)\left(\frac{W}{h} + 0.8\right)} \qquad (3.4)$$

$$L = \frac{1}{2f_r\sqrt{\varepsilon_{reff}}\sqrt{\mu_0\varepsilon_0}} - 2\Delta L \qquad (3.5)$$

where c is velocity of light in free space and ε_{reff} is effective dielectric constant.

The expressions given in equations 3.2 to 3.5 are used to calculate the L and W of a simple rectangular antenna for a resonant frequency of 2.32 GHz considering substrate parameters as $h = 3.175$ mm & $\varepsilon_r = 2.2$ and these values come out to be $L = 41.79$ mm and $W = 51.11$ mm resulting in an area of 2135.88 mm^2. The second iteration geometry of CRF antenna described above has a resonant frequency of 2.32 GHz and has dimensions of 30 mm x 37 mm resulting in an area of 1110 mm^2. Therefore, the second iteration CRF antenna has an area which is only 51.97% of 2135.88 mm^2. So, the CRF antenna results in a size reduction of 48.03%.

The above analysis shows that the designed CRF antenna has frequency reduction property which can be used for the design of miniaturized antennas. The drift in frequency is due to the fact that this geometry has multiple slots, which result in a larger current path leading to the lowering of resonant frequencies.

Tapered CRF Antenna

The scale ratio of the base and inserted rectangle in a specific iteration is 50% in the CRF antenna of the previous section. The fractal antenna of Fig. (**3.3**) is also simulated for different scale ratios, and the frequency reduction characteristic has been observed in all simulations. However, the main constraint of this fractal antenna geometry is that it has a very small bandwidth. Various approaches for the bandwidth enhancement of microstrip antennas have been proposed in recent years, *e.g.*, slotting [23, 24], tapering [25], the use of CPWs [26], *etc.*

In the presented work, diagonally opposite corners of the antenna are tapered to enhance the bandwidth. The fractal geometry of Fig. (**3.3 (c)**) with a scale ratio of 60% for the inner rectangles is selected for bandwidth enhancement. The taper dimensions are finalized as 4 mm x 4 mm using the trial and error method. Fig. (**3.5**) shows the values of various dimensions of the selected antenna.

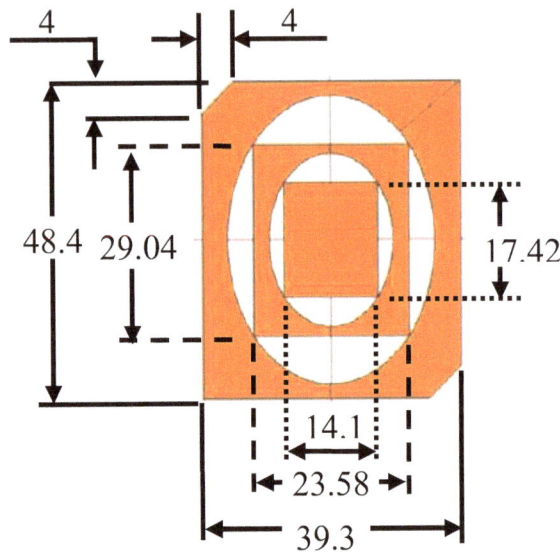

Fig. (3.5). Tapered CRF Antenna (All Dimensions in mm). (Reprinted from the Springer Nature: Neural Computing and Applications, BFO-ANN Ensemble Hybrid Algorithm to Design Compact Fractal Antenna for Rectenna System, Dhaliwal, B.S. and Pattnaik, S.S. © 2016).

The S_{11} results of the tapered and non-tapered second iteration CRF antenna given in Fig. (**3.6**) shows that the tapering significantly improves the bandwidth. The bandwidth of the non-tapered antenna is 38 MHz, and that of the tapered is 92.8 MHz resulting in an increase of about 144%. However, the tapering results in a shifting of resonant frequency from 1.856 GHz to 1.95 GHz, *i.e.*, a slight upwards shift in resonant frequency. The radiation patterns of the non-tapered and tapered antennas depict that the pattern shape is not disturbed due to tapering. However, the gain of tapered antenna has slightly reduced to 5.82 dBi from 5.97 dBi in the case of non-tapered [27].

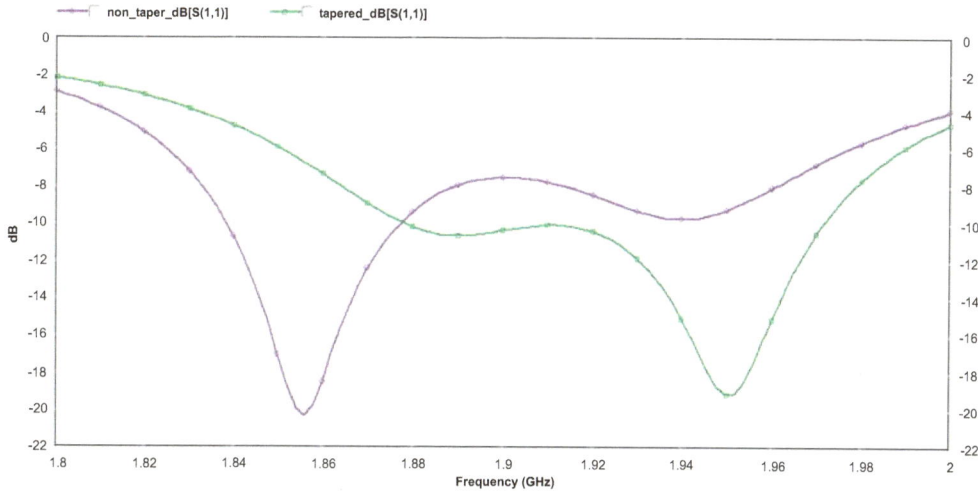

Fig. (3.6). S_{11} Results of Tapered and Non-Tapered CRF Antenna.

The tapered CRF described above has a resonant frequency of 1.95 GHz and has outer dimensions of 39.3 mm x 48.4 mm resulting in an area of 1902.12 mm². The expressions given in equations 3.2 to 3.5 are used to calculate the L and W of a simple rectangular antenna for the same resonant frequency, *i.e.*, 1.95 GHz considering substrate parameters as h = 3.175 mm & ε_r = 2.2 and these values come out to be L = 50.13 mm and W = 60.81 mm resulting in an area of 3048.41 mm². So, the tapered CRF antenna has an area of only 62.4% of the area of a simple rectangular antenna for the same frequency, so it results in a size reduction of 37.6%.

Miniaturized Crown Circular Fractal (CCF) Antenna

The CCF antenna geometry proposed here is based on a circular shape and is inspired by the fractal antenna of Ding *et al.* [28]. The development of the CCF antenna is shown in Fig. (3.7). The base geometry is a circle, as shown in Fig. (3.7(a)). The zeroth iteration geometry, shown in Fig. (3.7(b)), is obtained by cutting an ellipse from the base circular shape. To obtain the first iteration geometry shown in Fig. (3.7(c)), the reduced size copy of the zeroth iteration geometry is merged with the zeroth iteration such that the circumference of the inserted reduced shape touches the boundary of the elliptical slot. The same procedure can be repeated for the inner circular shape of the first iteration geometry to obtain the second iteration geometry, and similarly, further iterations can be designed.

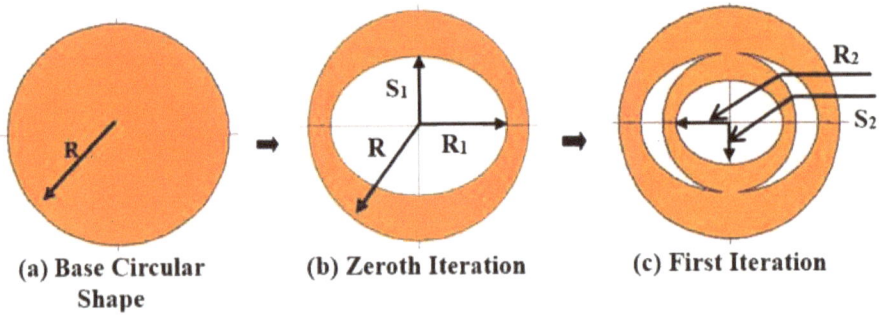

| (a) Base Circular Shape | (b) Zeroth Iteration | (c) First Iteration |

Fig. (3.7). CCF Antenna Geometries (Reprinted from the Springer Nature: Wireless Personal Communications, Development of PSO-ANN Ensemble Hybrid Algorithm and Its Application in Compact Crown Circular Fractal Patch Antenna Design, Dhaliwal, B.S. and Pattnaik, S.S. © 2017).

Fig. (**3.7(a)**) shows that the base geometry has one design variable, *i.e.*, the radius R of the circular antenna, Fig. (**3.7(b)**) depicts that the zeroth iteration shape has three variables: the radius R of the base circular antenna, the primary axis radius R_1 and the secondary axis radius S_1 of the elliptical slot, whereas the first iteration geometry has total five variables, *i.e.*, all the three variables (R, R_1, and S_1) of zeroth iteration and two more variables: the primary radius R_2 and the secondary radius S_2 of the inner elliptical slot as shown in Fig. (**3.7(c)**). In addition, the substrate parameters h & ε_r and the copper patch thickness t also affect the resonant frequency. Therefore, as shown in Fig. (**3.8(a)**), the resonant frequency of the first iteration of the proposed CCF antenna depends on seven parameters assuming $t \ll \lambda$. Consequently, its design for a desired frequency requires the optimal values of all these parameters. So, the design of the proposed CCF antenna is a multivariable problem [29].

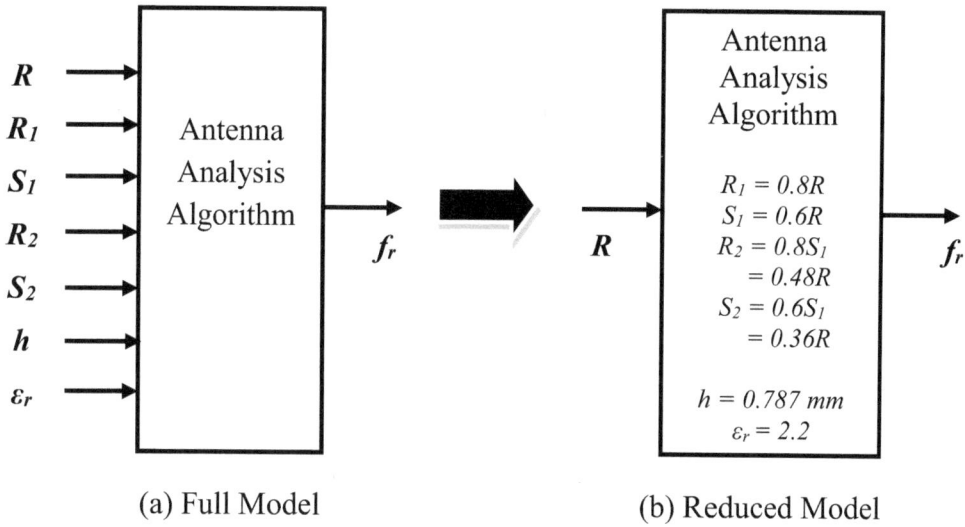

(a) Full Model　　　　　　　　　　　(b) Reduced Model

Fig. (3.8). Proposed CCF Antenna Analysis Model (Reprinted from the Springer Nature: Wireless Personal Communications, Development of PSO-ANN Ensemble Hybrid Algorithm and Its Application in Compact Crown Circular Fractal Patch Antenna Design, Dhaliwal, B.S. and Pattnaik, S.S. © 2017).

However, to reduce the number of design variables and hence, simplify the design procedure, the value of R_1 is taken as 80% of R and that of S_1 is taken as 60% of R. The radius of the inner circle of the first iteration shape is taken as equal to S_1 so that the circumference of the inserted reduced shape touches the boundary of the elliptical slot, and the value of R_2 is taken as 80% of S_1 and that of S_2 is taken as 60% of S_1. These assumptions make the number of design variables reduce to three, *i.e.*, the radius R of base circular shape, and the two substrate parameters h & ε_r. Further, in the presented design, the RT-Duroid substrate with $h = 0.787$ mm and $\varepsilon_r = 2.2$ is used. Therefore, as shown in Fig. (**3.8(b)**), the resonant frequency finally depends only on one parameter: the radius R of base circular shape. The other dimensions of the zeroth and first iteration geometries are calculated from the base radius R value using the above assumptions. The feed location of the proposed antenna is found by the trial-and-error approach; however, it can also be taken as another variable during antenna design.

In the presented work, the radius R of the base geometry of Fig. (**3.7(a)**) is taken as 10 mm. The zeroth iteration geometry of Fig. (**3.7(b)**) is achieved by cutting an ellipse from the base circular shape. As per the assumptions, the primary axis radius R_1 of the ellipse is 80% of R, *i.e.*, 8 mm and the secondary axis radius S_1 of the ellipse is 60% of R, *i.e.*, 6 mm. To obtain the dimensions of first iteration geometry shown in Fig. (**3.7(c)**), the zeroth iteration geometry is reduced by 60%, *i.e.*, outer radius of the reduced geometry is made equal to the secondary axis radius S_1 and then this reduced geometry is inserted in zeroth iteration geometry. As per the assumptions, the primary axis radius R_2 of the ellipse of the first iteration geometry is 80% of S_1, *i.e.*, 4.8 mm and the secondary axis radius S_2 of the ellipse of first iteration geometry is 60% of S_1 *i.e.*, 3.6 mm. The feeding points of base circular geometry, zeroth iteration, and first iteration are finalized by the trial-and-error method as (1.9, 1), (1.5, 6.2), and (1.7, 6.4) respectively with (0, 0) co-ordinate at the center of the geometry.

The proposed CCF antenna geometries are simulated using IE3D software and the analysis of simulated S_{11} plots of the zeroth and first iterations illustrate that the resonant frequency of the zeroth and first iteration antennas shift towards the lower end of the frequency scale as compared to the resonant frequency of circular antenna. This characteristic means that this geometry has frequency lowering properties which can be exploited for designing miniaturized antennas.

The radiation patterns of the CCF antenna depict that the CCF antenna has an omni-directional radiation pattern. The base geometry has a peak gain of 6.53 dBi. The peak gain of the first iteration geometry is equal to 6.03 dBi, whereas that of the zeroth iteration geometry is 5.99 dBi.

CONCLUSION

The chapter starts with an introduction to fractal geometry and its design procedure. Then the fractal antennas are defined, and the importance of self-similarity and space-filling properties in fractal antennas is described. The fractal antennas based on certain fractal shapes like the Sierpinski gasket, Sierpinski carpet, Koch curve, Hilbert curve, hexagonal, crown square, and rectangular Sierpinski carpet are discussed. The features of the antennas based on the above fractal geometries are highlighted, and it is seen that the fractal antennas have improved performances as compared to standard shapes of equivalent dimensions.

New fractal antennas developed in the presented research work are also described in this chapter. The miniaturized CRF antenna is designed on the rectangular base geometry, and it is seen that this antenna has size-reduction capability. A miniaturization of 48.03% is achieved for a 2.32 GHz frequency of operation. The bandwidth enhancement of this CRF antenna by tapering the opposite corner is discussed, and it is seen that tapering of corners resulted in an increase in bandwidth from 38 MHz to 92.8 MHz, an improvement of about 144%. However, the slight upward shift in resonant frequency is also observed due to tapering. The CCF antenna designed in the presented thesis work is based on a base circular shape. This fractal shape is developed on an RT-Duroid substrate, and it is seen that it also has miniaturization features. However, the gain of the CCF antenna developed on FR4 substrate is relatively less. Various parameters of this CCF geometry are interlinked to reduce the number of design variables.

ACKNOWLEDGMENT

The authors would like to thank IKG Punjab Technical University, Jalandhar, for providing the opportunity to publish this work.

DISCLOSURE

Part of this article has previously been published in the following articles:

• B. S. Dhaliwal and S. S. Pattnaik, "BFO–ANN ensemble hybrid algorithm to design a compact fractal antenna for rectenna system," Neural Comput. Appl., vol. 28, no. S1, pp. 917–928, 2017.

• B. S. Dhaliwal and S. S. Pattnaik, "Development of PSO-ANN ensemble hybrid algorithm and its application in compact crown circular fractal patch antenna design," *Wirel. Pers. Commun.*, vol. 96, no. 1, pp. 135–152, 2017.

REFERENCES

[1]   D.H. Werner, R.L. Haupt, and P.L. Werner, "Fractal antenna engineering: the theory and design of fractal antenna arrays", *IEEE Antennas Propag. Mag.*, vol. 41, no. 5, pp. 37-58, 1999.
      [http://dx.doi.org/10.1109/74.801513]

[2]   C. Puente, J. Romeu, R. Pous, X. Garcia, and F. Benitez, "Fractal multiband antenna based on the Sierpinski gasket", *Electron. Lett.*, vol. 32, no. 1, p. 1, 1996.
      [http://dx.doi.org/10.1049/el:19960033]

[3]   C.P. Baliarda, J. Romeu, and A. Cardama, "The Koch monopole: a small fractal antenna", *IEEE Trans. Antenn. Propag.*, vol. 48, no. 11, pp. 1773-1781, 2000.
      [http://dx.doi.org/10.1109/8.900236]

[4]   M.K.A. Rahim, M.Z.A.A. Aziz, and N. Abdullah, "Microstrip Sierpinski carpet antenna using transmission line feeding", *Proceedings of 2005 Asia-Pacific Microwave Conference*, pp. 1-4, 2005.
      [http://dx.doi.org/10.1109/APMC.2005.1606382]

[5]   J. Huang, F. Shan, J. She, and Z. Feng, "A novel small fractal patch antenna", *Proceedings of 2005 Asia-Pacific Microwave Conference Proceedings*, pp. 1-4, 2005.

[6]   K.J. Vinoy, J.K. Abraham, and V.K. Varadan, "On the relationship between fractal dimension and the performance of multi-resonant dipole antennas using koch curves", *IEEE Trans. Antenn. Propag.*, vol. 51, no. 9, pp. 2296-2303, 2003.
      [http://dx.doi.org/10.1109/TAP.2003.816352]

[7]   C. Puente-Baliarda, J. Romeu, R. Pous, and A. Cardama, "On the behavior of the Sierpinski multiband fractal antenna", *IEEE Trans. Antenn. Propag.*, vol. 46, no. 4, pp. 517-524, 1998.
      [http://dx.doi.org/10.1109/8.664115]

[8]   D.H. Wqrner, and S. Ganguly, "An overview of fractal antenna engineering research", *IEEE Antennas Propag. Mag.*, vol. 45, no. 1, pp. 38-57, 2003.
      [http://dx.doi.org/10.1109/MAP.2003.1189650]

[9]   J.P. Gianvittorio, and Y. Rahmat-Samii, "Fractal antennas: a novel antenna miniaturization technique, and applications", *IEEE Antennas Propag. Mag.*, vol. 44, no. 1, pp. 20-36, 2002.
      [http://dx.doi.org/10.1109/74.997888]

[10] B. Panoutsopoulos, "Printed circuit fractal antennas", *Proceedings of 2003 IEEE International Conference on Consumer Electronics,* pp. 288-289, 2003.

[11] T. Tiehong, and Z. Zheng, "A novel multiband antenna: fractal antenna", *Proceedings of International Conference on Communication Technology Proceedings,* pp. 1907-1910, 2003.

[12] M.F.M. Yusof, I.P. Pohan, M. Esa, N.A. Murad, and Y.E. Chuan, "Stacked square fractal antenna with improved bandwidth for wireless local area network access point", *International RF and Microwave Conference,* pp. 228-232, 2006.
[http://dx.doi.org/10.1109/RFM.2006.331075]

[13] N. Song, K. Chin, D. Boon Liang, and M. Anyi, "Design of broadband dual-frequency microstrip patch antenna with modified Sierpinski fractal geometry", *Proceedings of 2006 10th IEEE Singapore International Conference on Communication Systems,* pp. 1-5, 2006.
[http://dx.doi.org/10.1109/ICCS.2006.301376]

[14] R. Azaro, G. Boato, M. Donelli, A. Massa, and E. Zeni, "Design of a prefractal monopolar antenna for 3.4-3.6 GHz WI-max band portable devices", *IEEE Antennas Wirel. Propag. Lett.,* vol. 5, pp. 116-119, 2006.
[http://dx.doi.org/10.1109/LAWP.2006.872427]

[15] S. Wong, B.L. Ooi, P.S. Kooi, and M.S. Leong, "An improved microstrip Sierpinski carpet antenna", *Proceedings of 2001 Asia-Pacific Microwave Conference,* pp. 483-486, 2001.

[16] C. Puente, J. Romeu, R. Pous, J. Ramis, and A. Hijazo, "Small but long Koch fractal monopole", *Electron. Lett.,* vol. 34, no. 1, p. 9, 1998.
[http://dx.doi.org/10.1049/el:19980114]

[17] P.W. Tang, and P.F. Wahid, "Hexagonal fractal multiband antenna", *IEEE Antennas Wirel. Propag. Lett.,* vol. 3, pp. 111-112, 2004.
[http://dx.doi.org/10.1109/LAWP.2004.829989]

[18] P. Dehkhoda, and A. Tavakoli, "A crown square microstrip fractal antenna", *Proceedings of IEEE Antennas and Propagation Society Symposium,* pp. 2396-2399, 2004.
[http://dx.doi.org/10.1109/APS.2004.1331855]

[19] N.L. Nhlengethwa, and P. Kumar, "Fractal microstrip patch antennas for dual-band and triple-band wireless applications", *Int. J. Smart Sensing Intell. Syst.,* vol. 14, no. 1, pp. 1-9, 2021.
[http://dx.doi.org/10.21307/ijssis-2021-007]

[20] Wen-Ling Chen, Guang-Ming Wang, and Chen-Xin Zhang, "Small-size microstrip patch antennas combining Koch and Sierpinski fractal-shapes", *IEEE Antennas Wirel. Propag. Lett.,* vol. 7, pp. 738-741, 2008.
[http://dx.doi.org/10.1109/LAWP.2008.2002808]

[21] J. Anguera, A. Andújar, J. Jayasinghe, V. S. Chakravarthy, P. S. R. Chowdary, J. L. Pijoan, T. Ali, and C. Cattani, "Fractal antennas: an historical perspective", *Fractal Fract.,* vol. 4, no. 1, p. 3(1-26), 2020.

[22] C.A. Balanis, *Antenna theory: Analysis and design.* 3rd ed. Wiley-Blackwell: Chichester, England, 2005.

[23] W.-L. Chen, G.-M. Wang, and C.-X. Zhang, "Bandwidth enhancement of a microstrip-line-fed printed wide-slot antenna with a fractal-shaped slot", *IEEE Trans. Antenn. Propag.,* vol. 57, no. 7, pp. 2176-2179, 2009.
[http://dx.doi.org/10.1109/TAP.2009.2021974]

[24] J.-Y. Jan, and Jia , "Bandwidth enhancement of a printed wide-slot antenna with a rotated slot", *IEEE Trans. Antennas Propag,* vol. 53, no. 6, pp. 2111-2114, 2005.

[25] R. Ghatak, S. Chatterjee, and D.R. Poddar, "Wideband fractal shaped slot antenna for X-band application", *Electron. Lett.,* vol. 48, no. 4, p. 198, 2012.
[http://dx.doi.org/10.1049/el.2011.3483]

[26]   M. Naghshvarian Jahromi, A. Falahati, and R.M. Edwards, "Bandwidth and impedance-matching enhancement of fractal monopole antennas using compact grounded coFEM", *IEEE Trans. Antenn. Propag.,* vol. 59, no. 7, pp. 2480-2487, 2011.
[http://dx.doi.org/10.1109/TAP.2011.2152321]

[27]   B.S. Dhaliwal, and S.S. Pattnaik, "Performance comparison of bio-inspired optimization algorithms for Sierpinski gasket fractal antenna design", *Neural Comput. Appl.,* vol. 27, no. 3, pp. 585-592, 2016.
[http://dx.doi.org/10.1007/s00521-015-1879-y]

[28]   M. Ding, R. Jin, J. Geng, Q. Wu, and W. Wang, "Design of a CPW-fed ultra wideband crown circular fractal antenna", *Proceedings of 2006 IEEE Antennas and Propagation Society International Symposium,* pp. 2049-2052, 2006.
[http://dx.doi.org/10.1109/APS.2006.1710983]

[29]   B.S. Dhaliwal, and S.S. Pattnaik, "Development of PSO-ANN ensemble hybrid algorithm and its application in compact crown circular fractal patch antenna design", *Wirel. Pers. Commun.,* vol. 96, no. 1, pp. 135-152, 2017.
[http://dx.doi.org/10.1007/s11277-017-4157-8]

CHAPTER 4

Development of ANN Models for the Design of Fractal Antennas

Abstract: In this chapter, the development of ANN models for the design of proposed fractal antennas is explained. The various parameters of the fractal antennas selected for ANN models are described. The ANN models are designed using feed-forward neural networks, namely MLPNN, RBFNN and GRNN. The performance comparison of different ANN models on the basis of different performance measures is also given. The design of ANN ensemble models for fractal antennas is introduced, and different techniques for developing ANN ensemble models are also discussed in this chapter.

Keywords: ANN, ANN ensemble, Crown fractal antenna, Fractal antenna, Miniaturized antenna.

INTRODUCTION

ANNs have been used for the design of microstrip patch antennas by a number of researchers. ANNs are applied by Mishra and Patnaik [1] and Turker *et al*. [2] to design rectangular patch antennas. ANN coupled with GA is used by Panda *et al*. [3] and Khuntia *et al*. [4] for designing rectangular microstrip antennas on thick substrates. Kumar *et al*. [5] presented the use of ANN for parameter estimation of a multislotted rectangular microstrip antenna. The use of a PSO-driven RBFNN to design an equilateral triangular microstrip antenna is presented by Chintakindi *et al*. [6]. A circularly-polarized square microstrip antenna is designed by Wang *et al*. [7]. ANN is used by Siakavara [8] to design a circular microstrip ring antenna for multi-frequency operation. The application of ANN for designing a rectangular microstrip ring antenna with proximity-coupled feed is proposed by Manh *et al*. [9]. An application of ANN for the design of an inset-fed rectangular microstrip antenna is discussed by Vilovic *et al*. [10]. The design of a circular microstrip antenna using ANN is presented by Gultekin *et al*. [11]. Bose and Gupta [12] have presented an ANN model based on RBF and a back-propagation algorithm to design aperture-coupled microstrip antennas.

ANNs are also used to estimate the resonant frequencies of microstrip antennas. Devi *et al*. [13] and Khuntia *et al*. [14] used ANNs to calculate the resonant frequency of different rectangular microstrip patch antennas. Pattnaik *et al.* [15] presented an ANN model to compute the resonant frequency of a single-shorting post-tunable rectangular microstrip-patch antenna. The resonant frequency of a rectangular microstrip antenna with and without an air gap is calculated using ANN by Tighilt *et al.* [16]. ANN models are used by Can *et al.* [17] for estimating the resonant frequencies of a dual-band equilateral triangular microstrip antenna. Yu-Bo *et al.* [18] presented the use of an ANN ensemble for modeling the resonant frequency of a rectangular microstrip antenna. The ANN models have also found a number of applications in estimating parameters other than the resonant frequencies. Hettak and Delisle [19] used ANN to calculate the radiation efficiency of a rectangular microstrip patch antenna. Neog *et al.* [20] designed a tunnel-based ANN model for the parameter calculation of a wideband microstrip antenna. The input impedance of a loop antenna is predicted by Kim *et al.* [21] by using the ANN models. Panda *et al.* [22] used ANN to speed up FDTD calculations to evaluate the input impedance of a stacked microstrip patch antenna. The feed point of a circular microstrip antenna is estimated with the help of the RBFNN model by Vilovic and Burum [23]. An application of the RBFNN model to analyze the bandwidth of a slot-loaded triple-band patch antenna is proposed by Aneesh *et al.* [24]. The use of ANNs for the analysis and design of antenna arrays has been reported by various researchers. Patnaik *et al.* [25] proposed an ANN-based approach to locate the faulty elements in antenna arrays. The optimization of energy parameters of antenna arrays using the ANN is proposed by Bashly and Popovskii [26]. The side-lobe reduction of an antenna array is achieved by Lee *et al.* [27] using an ANN-based automatic converging scheme. Vakula and Sarma [28] proposed the use of the RBFNN model for the diagnosis of planar antenna arrays from far-field radiation patterns. Zaharis *et al.* [29] presented ANN models for adaptive beamforming of antenna arrays. The use of RBFNN for directivity estimations of arrays of short dipoles is presented by Mishra *et al.* [30].

The above applications of ANNs in microstrip antennas' analysis and design show that the ANNs are very suitable in this field. The applications of ANNs for the parameter estimation of fractal antennas are developed in the presented research work and are described in this chapter.

DEVELOPMENT OF ANN MODELS FOR FRACTAL ANTENNAS

As compared to the traditional patch antennas, the fractal antenna patch shapes are complex, and the mathematical formulas for the analysis and design of these antennas do not exist. ANNs are suitable choices for fractal antennas because development of analytical methods is challenging for new structures, numerical modeling methods are computationally expensive, and empirical models have limited range and accuracy [31]. The following sections describe the ANN models developed for the analysis and design of fractal antennas designed in this research work. The input and output parameters of the ANN model depend on the fractal shape as well as on the relationships to be modeled. Three feed-forward ANN types, namely MLPNNs, RBFNNs, and GRNNs, are considered in the presented work.

ANN Model for Analysis of SGMF Antenna

The SGMF antenna has been explored most widely than any other fractal antenna since its presentation in 1998 by Puente-Baliarda *et al.* [32]. The development of the first four iterations is shown in Fig. (**4.1**). This antenna is a multiband antenna, and the number of bands depends on the number of iterations *n* [32]. The zeroth iteration (base triangular shape) has only one resonance frequency. The first iteration antenna has two resonant frequencies and so on. The side length *s* of the antenna, ε_r, and *h* of the substrate also affect the resonant frequencies. So, the resonant frequencies f_r, depend upon the values of ε_r, *h*, *s* and number of iterations *n*.

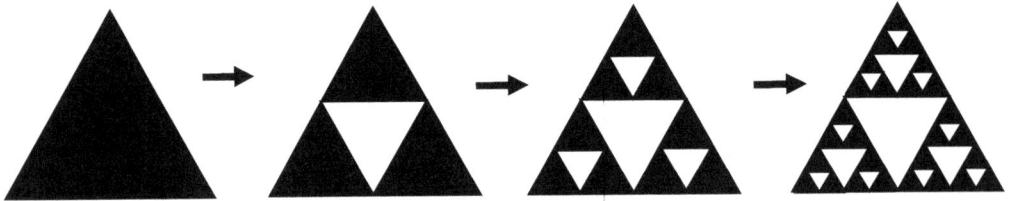

Fig. (4.1). First Four Iterations of SGMF Antenna (Reprinted from the Springer Nature: Neural Computing and Applications, Performance Comparison of Bio-Inspired Optimization Algorithms for Sierpinski Gasket Fractal Antenna Design, Dhaliwal, B.S. and Pattnaik, S.S. © 2016).

A closed-form expression for estimating the resonant frequency f_r of this antenna is also proposed by Puente-Baliarda *et al.* [32], which is first modified by Song *et al.* [33] and then by Mishra *et al.* [34]. The latest expression is given below in equation (4.1):

$$f_r = \begin{cases} (0.15345 + 0.34\rho x)\frac{c}{H_e}\,(\xi^{-1})^n & for\ n = 0 \\ 0.26\frac{c}{H_e}\delta^n & for\ n > 0 \end{cases}$$

$$(4.1)$$

where H_e is effective height of the largest Sierpinski gasket defined by $H_e = \frac{\sqrt{3}s_e}{2}$,

s_e being the effective side-length of the gasket and it is given by $s_e = s + \dfrac{h}{\sqrt{\varepsilon_r}}$, c is speed of light, δ is scale ratio and its value is *2* for antenna under consideration, ξ is defined as equal to *1/δ*, ρ is equal to ξ - *0.230735*, and x is defined as *0* for *n = 0*.

The ANN models for this antenna are developed as described below, and the ANN results are compared with theoretical results calculated using equation (4.1) and the experimental results. The ANN model developed for the SGMF antenna is shown in Fig. (**4.2**). The input parameters are: *s*, *n*, ε_r, and *h* and the output of the ANN model is the corresponding f_r. The data set used for the training of the ANN model consists of values of input parameters and corresponding output values [35].

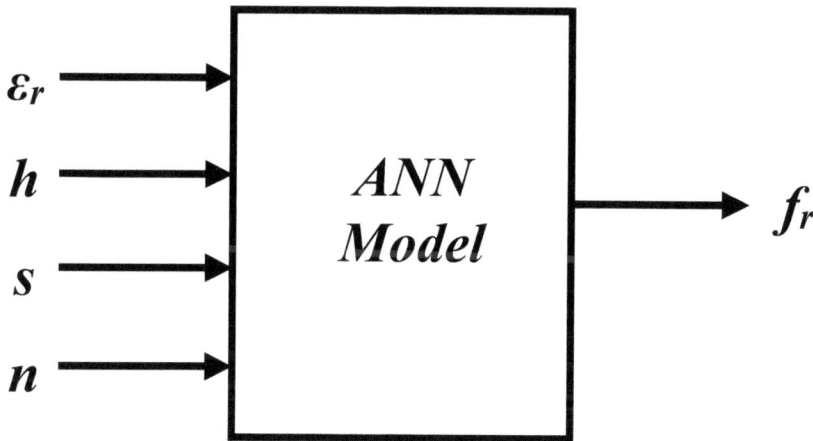

Fig. (4.2). ANN Model for SGMF Antenna [35].

The ANN model of Fig. (**4.2**) has been implemented using three types of ANNs: MLPNN, RBFNN, and GRNN. The architecture used for MLPNN consists of 4 input neurons, 15 hidden layer neurons and 1 neuron in the output layer. The training function used is *trainlm*, and the value of the learning rate is 0.2. For RBFNN, the number of neurons in the hidden layer is 45, *i.e.*, equal to the number of sets of training data employed for training. The value of the spread constant has been selected as 1.05. In the GRNN model also, 45 neurons are used in the hidden layer. For the training of GRNN, the value of the spread constant has been selected as 0.85. All three ANN models described above are trained and tested using the test data set. Then these models are utilized to estimate the resonant frequencies (f_0, f_1, f_2, f_3, f_4) of a fourth iteration Sierpinski gasket antenna with parameters ε_r = 2.5, s = 102.7683 mm, and h = 1.588 mm. ANN results are compared with the experimental and theoretical results and are shown in Table **4.1** which depicts that the error values are very small for all the models.

Table 4.1. Comparison of the Results of ANN Models for SGMF Antenna [35].

Resonant Frequency f_r	Experi-mental Results (GHz) [32]	Theoretical Results [34]		MLPNN Results [35]		RBFNN Results [35]		GRNN Results [35]	
		Output (GHz)	Abso-lute Error	Output (GHz)	Abso-lute Error	Output (GHz)	Abso-lute Error	Output (GHz)	Abso-lute Error
f_0	0.520	0.5140	0.0060	0.5122	0.0078	0.5123	0.0077	0.8922	0.3722
f_1	1.740	1.7420	0.0020	1.7362	0.0038	1.7360	0.0040	1.9097	0.1697
f_2	3.510	3.4840	0.0260	3.4712	0.0388	3.4717	0.0383	3.9282	0.4182
f_3	6.950	6.9680	0.0180	6.9437	0.0063	6.9434	0.0066	7.6243	0.6743
f_4	13.89	13.9340	0.0440	13.8865	0.0035	13.8867	0.0033	11.8325	2.0575

Two different performance measures are used to evaluate the developed ANN models. These are (i) Mean Absolute Error (*MAE*), which is a criterion of the effectiveness of the training of ANN, and (ii) coefficient of correlation (C_R), which gives an idea about the linear relationship developed by ANN model. The smaller value of *MAE* means better performance, and in case of C_R, the good performance is indicated by a value of C_R near to 1. The values of these performance measures for the models trained above are given in Table **4.2**.

Table 4.2. Performance Measures for Trained ANN Models [35].

Performance Measure	Theoretical Results [34]	MLPNN Results [35]	RBFNN Results [35]	GRNN Results [35]
MAE	0.0192	0.0120	0.0120	0.7384
C_R	1.0000	1.0000	1.0000	0.9899

The comparison between the three models on the basis of *MAE* shows that the MLPNN and RBFNN have the smallest error values and are even better than the theoretical results. It may be seen that the values of C_R are equal to 1 for MLPNN, RBFNN, and theoretical results, and it is sufficiently high for GRNN, indicating satisfactory performance by all models. When both performance criteria are considered together, the most satisfactory model is RBFNN, which performs better than MLPNN, and GRNN and is even more accurate than the theoretical method in predicting resonant frequencies. Also due to the fast adaptive properties of ANN, the simulation time required is very less. Thus, the ANN approach to

Sierpinski fractal antenna analysis is a low-cost, accurate and computationally fast approach.

Parameter Estimation of CRF Antenna using ANN Models

ANN models for parameter estimation of CRF antenna shown in Fig. (**4.3**) are developed and compared to find the suitable type for this antenna. Fig. (**4.3**) shows the first two iterations of the CRF antenna, which are similar to the CRF antenna of Fig (**3.3**) of Chapter 3, except that the scale ratio is 60% for this antenna. The substrate of $h = 3.175$ mm, $\varepsilon_r = 2.2$, and tan δ of 0.0009 are used.

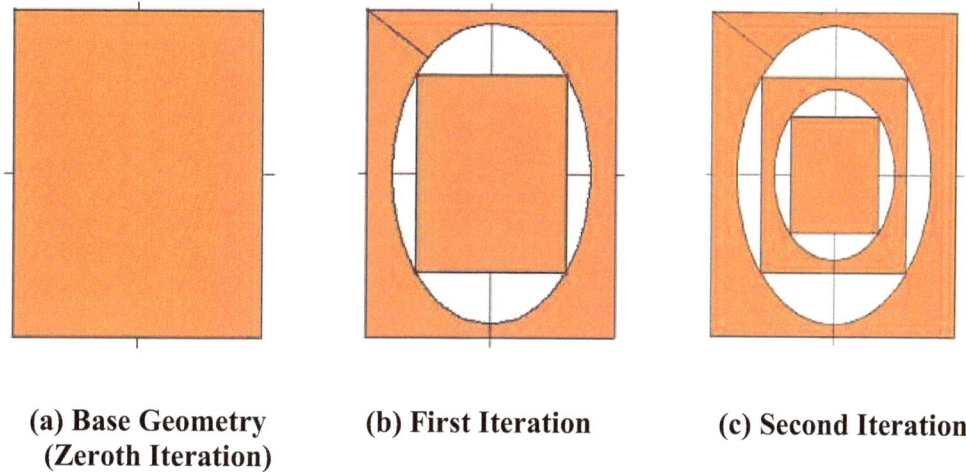

| (a) Base Geometry (Zeroth Iteration) | (b) First Iteration | (c) Second Iteration |

Fig. (4.3). CRF Antenna © 2012. IEEE, Reprinted with Permission from Dhaliwal, B.S. and Pattnaik, S.S., "Performance Evaluation of Artificial Neural Networks in Microstrip Fractal Antenna Parameter Estimation", Proceedings of IEEE International Conference on Communication Systems, Singapore, 2012, pp. 135-139.

As shown in Fig. (**4.4**), the ANN model has three inputs: feed location co-ordinates (x_i, y_i) and iteration number (n). The corresponding resonant frequency (f_r), return loss (S_{11}) and gain (G) are taken as outputs. The location of the feed point is very important in antenna performance. The feed point must be located at that point on the patch where the input impedance is 50 ohms for the resonant frequency. But it is not an easy task to achieve in the case of fractal antennas because of the complex geometry of different iterations. The proposed ANN model shows the effect of feed location in different iterations on the antenna parameters, so these models can be used to finalize the feed point in different locations. Three ANN types: the MLPNN, RBFNN, and GRNN, are used to implement the ANN block diagram of Fig. (**4.4**). A data set is obtained using IE3D software for 75 different values of feed locations of CRF antenna of different iterations with outer dimensions 40.3 mm and 49.4 mm. Out of these 75 values, 55 are used to train the ANN models. For the training of the first type, *i.e.*, MLPNN, 3 input neurons and 40 neurons in hidden layer, and 3 output neurons are used. The *trainlm* function is used as the training function and the learning rate is selected as 0.25. For the training of RBFNN and GRNN, the value of the spread constant has been selected as 1.05 and 0.85, respectively [36].

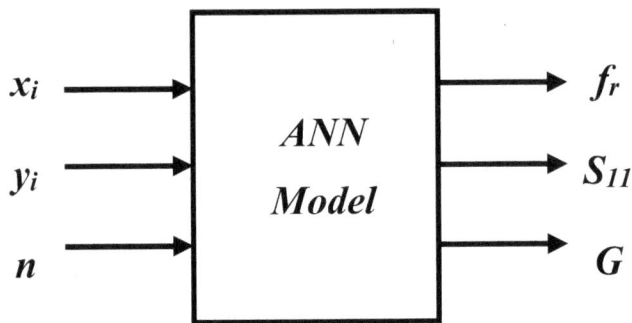

Fig. (4.4). ANN Model for CRF Antenna © 2012. IEEE, Reprinted with Permission from Dhaliwal, B.S. and Pattnaik, S.S., "Performance Evaluation of Artificial Neural Networks in Microstrip Fractal Antenna Parameter Estimation", Proceedings of IEEE International Conference on Communication Systems, Singapore, 2012, pp. 135-139.

The trained ANN models are tested by taking the remaining 20 values of the data set, and ANN results are compared with the IE3D simulation results, as shown in Tables **4.3**, **4.4**, and **4.5**. All values of x_i and y_i are written by taking the centre of proposed geometries at the origin of the co-ordinate axis.

Table 4.3. Comparison of ANN Results for First Output Parameter *i.e.*, f_r.

S No.	ANN Inputs			IE3D Output	MLPNN Output		RBFNN Output		GRNN Output	
	x_i (mm)	y_i (mm)	n	f_r (GHz)	f_r (GHz)	Absolute Error	f_r (GHz)	Absolute Error	f_r (GHz)	Absolute Error
1	10.00	4.00	0	2.359	2.4608	0.1018	2.1729	0.1861	2.3500	0.0090
2	14.25	17.65	0	2.388	2.3828	0.0052	1.8647	0.5233	2.2788	0.1092
3	8.00	23.65	0	2.33	2.2062	0.1238	1.8476	0.4824	2.3200	0.0100
4	-19.12	5.96	0	2.42	1.4826	0.9374	1.8483	0.5717	1.8500	0.5700
5	-8.00	14.00	0	2.334	2.3424	0.0084	1.8868	0.4472	2.3151	0.0189
6	8.25	8.75	0	2.34	2.3669	0.0269	2.1487	0.1913	2.0252	0.3148
7	9.00	11.00	1	1.87	1.8793	0.0093	1.8569	0.0131	1.8439	0.0261
8	-14.00	-16.00	1	1.90	1.8909	0.0091	1.8706	0.0294	2.3800	0.4800
9	19.00	4.00	1	1.85	-0.0412	1.8912	1.8476	0.0024	1.8602	0.0102

(Table 4.3) cont.....

10	-9.50	9.10	1	1.85	1.8635	0.0135	1.8573	0.0073	1.8503	0.0003
11	10.10	-8.10	1	1.86	1.7529	0.1071	1.8477	0.0123	1.8800	0.0200
12	8.36	9.35	1	1.86	1.8640	0.0040	1.8585	0.0015	1.8621	0.0021
13	-15.10	17.50	1	1.90	2.0510	0.1510	1.8476	0.0524	2.3329	0.4329
14	8.10	24.10	2	1.89	1.9042	0.0142	1.8476	0.0424	2.3200	0.4300
15	9.75	-12.56	2	1.849	0.9075	0.9415	1.8499	0.0009	1.8880	0.0390
16	10.00	12.00	2	1.84	1.8792	0.0392	1.8471	0.0071	1.8432	0.0032
17	-6.40	8.20	2	1.819	1.9055	0.0865	1.8411	0.0221	1.8403	0.0213
18	-10.41	11.90	2	1.839	1.7985	0.0405	1.8436	0.0046	1.8394	0.0004
19	16.34	13.59	2	1.84	1.8429	0.0029	1.8485	0.0085	1.8485	0.0085
20	-8.25	11.50	2	1.84	1.8407	0.0007	1.8436	0.0036	1.8401	0.0001
					MAE = 0.1799		**MAE = 0.1305**		**MAE = 0.1253**	

*© 2012. IEEE, Reprinted with Permission from Dhaliwal, B.S. and Pattnaik, S.S., "Performance Evaluation of Artificial Neural Networks in Microstrip Fractal Antenna Parameter Estimation", Proceedings of IEEE International Conference on Communication Systems, Singapore, 2012, pp. 135-139.

Table 4.4. Comparison of ANN Results for second Output Parameter *i.e.* S_{11}.

S No.	ANN Inputs			IE3D Output	MLPNN Output		RBFNN Output		GRNN Output	
	x_i (mm)	y_i (mm)	n	S_{11} (dB)	S_{11} (dB)	Absolute Error	S_{11} (dB)	Absolute Error	S_{11} (dB)	Absolute Error
1	10.00	4.00	0	-20.23	-18.3493	1.8807	-19.1925	1.0375	-22.0600	1.8300
2	14.25	17.65	0	-8.5	-19.1142	10.6142	-13.7057	5.2057	-10.8933	2.3933
3	8.00	23.65	0	-31.76	-26.0084	5.7526	-13.9307	17.8303	-16.7200	15.0410
4	-19.12	5.96	0	-8.1	-13.5467	5.4467	-10.6520	2.5520	-2.2887	5.8113
5	-8.00	14.00	0	-27.3	-28.3955	1.0955	-15.2727	12.0273	-30.2032	2.9032
6	8.25	8.75	0	-31.04	-34.6884	3.6484	-30.1407	0.8993	-29.3994	1.6406
7	9.00	11.00	1	-16.54	-18.9352	2.3952	-11.7293	4.8107	-11.4906	5.0494
8	-14.00	-16.00	1	-14.82	-11.9533	2.8667	-13.7052	1.1148	-8.7002	6.1198
9	19.00	4.00	1	-1.62	8.7142	10.3342	-13.9306	12.3106	-4.1521	2.5321
10	-9.50	9.10	1	-1.61	-20.8807	19.2707	-19.8554	18.2454	-24.4936	22.884
11	10.10	-8.10	1	-19.72	-20.7237	1.0037	-13.9454	5.7746	-16.9700	2.7500
12	8.36	9.35	1	-27.9	-33.8309	5.9309	-38.1606	10.2606	-29.7923	1.8923
13	-15.10	17.50	1	-12.88	2.5020	15.3820	-13.9307	1.0507	-30.7975	17.917
14	8.10	24.10	2	-8.8	3.8331	12.6331	-13.9307	5.1307	-16.7200	7.9200
15	9.75	-12.56	2	-16.97	-4.1809	12.7891	-14.0817	2.8883	-17.0714	0.1014
16	10.00	12.00	2	-17.52	-15.8059	1.7141	-13.5479	3.9721	-14.1993	3.3207

(Table 4.4) cont.....

17	-6.40	8.20	2	-15.2	7.9577	23.1577	10.0225	25.2225	13.4418	28.6418
18	-10.41	11.90	2	-19.06	-25.5833	6.5233	-15.6588	3.4012	-15.9897	3.0703
19	16.34	13.59	2	-9.5	-13.6036	4.1036	-13.8633	4.3633	-11.6534	2.1534
20	-8.25	11.50	2	-14.04	-15.4322	1.3922	-12.2945	1.7455	-14.6766	0.6366
					MAE = 7.3967		MAE = 6.9921		MAE = 6.7304	

Table 4.5. Comparison of ANN Results for Third Output Parameter *i.e. G.*

S No.	ANN Inputs			IE3D Output	MLPNN Output		RBFNN Output		GRNN Output	
	x_i (mm)	y_i (mm)	n	G (dBi)	G (dBi)	Absolute Error	G (dBi)	Absolute Error	G (dBi)	Absolute Error
1	10.00	4.00	0	7.03	6.3183	0.7117	6.9284	0.1016	6.7800	0.2500
2	14.25	17.65	0	5.07	5.1728	0.1028	7.0840	2.0140	5.8205	0.7505
3	8.00	23.65	0	7.03	11.9621	4.9321	7.2013	0.1713	6.7700	0.2600
4	-19.12	5.96	0	4.95	9.2194	4.2694	5.8498	0.8998	2.4187	2.5313
5	-8.00	14.00	0	5.76	5.0172	0.7428	7.1704	1.4104	6.8755	1.1155
6	8.25	8.75	0	7.07	7.1613	0.0913	6.8981	0.1719	6.4342	0.6358
7	9.00	11.00	1	5.9	3.6949	2.2051	5.8603	0.0397	3.5574	2.3426
8	-14.00	-16.00	1	5.93	6.1162	0.1800	7.1137	1.1837	5.1701	0.7599
9	19.00	4.00	1	1.61	-5.0891	6.6991	7.2013	5.5913	4.2005	2.5905
10	-9.50	9.10	1	1.85	6.5475	4.6975	6.7313	4.8813	6.6246	4.7746
11	10.10	-8.10	1	6.2	3.4636	2.7364	7.1982	0.9982	6.5700	0.3700
12	8.36	9.35	1	6.4	5.8442	0.5558	6.1692	0.2308	6.1427	0.2573
13	-15.10	17.50	1	6.47	19.0936	12.6236	7.2013	0.7313	6.8999	0.4299
14	8.10	24.10	2	6	9.7105	3.7105	7.2013	1.2013	6.7700	0.7700
15	9.75	-12.56	2	5.8	1.9407	3.8593	7.1724	1.3724	6.5513	0.7513
16	10.00	12.00	2	5.99	6.4142	0.4242	6.5508	0.5608	4.3611	1.6289
17	-6.40	8.20	2	5.93	3.7750	2.1550	4.7268	1.2032	4.4531	1.4769
18	-10.41	11.90	2	6.53	9.4005	2.8705	6.9138	0.3838	6.2723	0.2577
19	16.34	13.59	2	6.12	10.1683	4.0483	7.1738	1.0538	6.2834	0.1634
20	-8.25	11.50	2	6	5.7766	0.2234	5.5464	0.4536	6.0231	0.0231
					MAE = 2.8922		MAE = 1.2327		MAE = 1.1070	

All the tables above clearly show that the MAE is minimum in the case of GRNN for all outputs. The time taken by the networks for training is calculated to be equal to 17.89 seconds, 0.137 seconds, and 0.036 seconds for MLPNN, RBFNN, and GRNN, respectively. Clearly, the GRNN takes very less time as compared to other types. Therefore, the GRNN model is the best choice for such applications.

ANN ENSEMBLE MODELS FOR FRACTAL ANTENNAS

As the fractal antennas have complex shapes, so, it is logical to use an ANN ensemble model instead of a single ANN model for modeling the behaviour of fractal antennas.

ANN Ensemble Model for Tapered CRF Antenna

This section describes an ANN ensemble model which estimates the resonant frequency for the given dimensions of the tapered CRF antenna described in Chapter 3 and shown below in Fig. (**4.5**). As discussed in Chapter 3, the f_r of the tapered CRF antenna depends upon the dimensions of the base rectangle. Therefore, the ANN ensemble model, shown in Fig. (**4.6**), has two inputs: the length L and the width W of the base rectangle of the tapered CRF antenna, and one output: the corresponding resonant frequency f_r. Therefore, the ANN ensemble estimates the values of f_r for the given values of L and W. The design of ANN ensembles requires the decision about the number of individual ANN models to be connected in parallel & their structure, the method of combining the output of individual ANN models, and the training and testing of the data set [37].

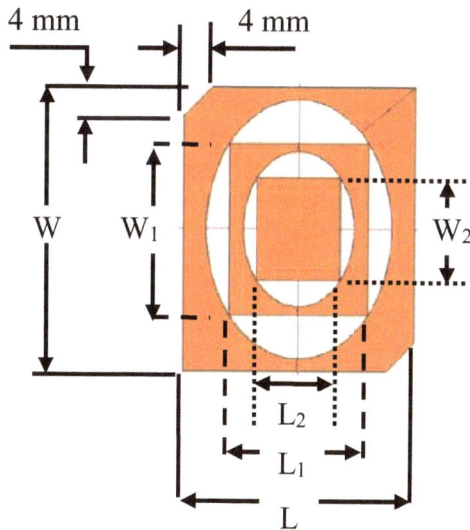

Fig. (4.5). Tapered CRF Antenna [38].

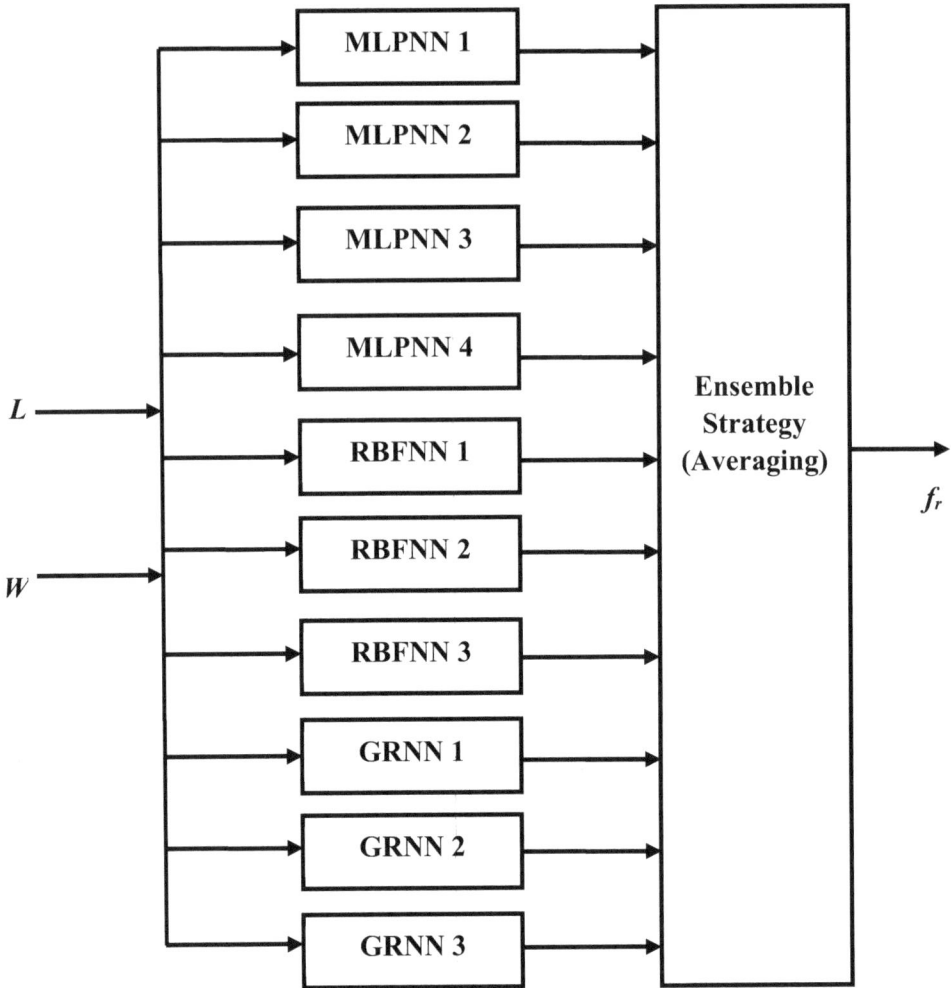

Fig. (4.6). Block Diagram of ANN Ensemble for Tapered CRF Antenna (Reprinted from the Springer Nature: Neural Computing and Applications, BFO-ANN Ensemble Hybrid Algorithm to Design Compact Fractal Antenna for Rectenna System, Dhaliwal, B.S. and Pattnaik, S.S. © 2016).

As shown in Fig. (4.6), ten different ANN models, which include four MLPNNs trained with a back-propagation algorithm, three RBFNNs, and three GRNNs, are used to implement the ensemble. The same type of ANN models differs in their internal structures, as shown in Table **4.6**. The outputs of the individual ANN models are combined by a simple averaging method to generate the final ensemble output [38].

The data set used for the training and testing is generated by using the IE3D simulator. Forty different values of L in the range 30 mm to 38 mm and W in the range 35 mm to 50 mm are randomly generated. The tapered CRF antenna of these dimensions is simulated using the IE3D simulator and corresponding resonant frequencies are calculated. The taper dimensions are kept as 4 mm x 4 mm for all simulations. Out of these forty samples, thirty samples are used for training and the remaining ten samples are used for testing.

The training of the individual ANN models is an important issue as the training strategy influences the diversity and accuracy of individual ANN models in the ensemble [37]. There are two different groups of algorithms to generate diverse training sets: (i) the algorithms that take into account the performance of the previous classifiers to adaptively change the distribution of the training set, like the boosting algorithm, and (ii) the algorithms that do not adapt the distribution like the bagging algorithm [39].

Table 4.6. Details of ANN Models Employed in Ensemble Implementation.

S. No.	Network Name	Network Type	No. of Hidden Layers	No. of Hidden Neurons	Other Parameters
1	MLPNN 1	MLPNN	2	20, 15	Training Function = *Trainlm* Learning Rate = 0.2 Performance Goal = 10^{-6}
2	MLPNN 2	MLPNN	1	25	Training Function = *Trainlm* Learning Rate = 0.15 Performance Goal = 10^{-6}
3	MLPNN 3	MLPNN	3	10, 15, 10	Training Function = *Trainbfg* Learning Rate = 0.1 Performance Goal = 10^{-5}
4	MLPNN 4	MLPNN	2	15, 15	Training Function = *Traincgf* Learning Rate = 0.15 Performance goal = 10^{-5}
5	RBFNN 1	RBFNN	1	25	Spread Constant = 0.25
6	RBFNN 2	RBFNN	1	22	Spread Constant = 0.55
7	RBFNN 3	RBFNN	1	20	Spread Constant = 1.4
8	GRNN 1	GRNN	2	22, 2	Spread Constant = 0.85
9	GRNN 2	GRNN	2	25, 2	Spread Constant = 0.7
10	GRNN 3	GRNN	2	20, 2	Spread Constant = 0.9

*Reprinted from the Springer Nature: Neural Computing and Applications, BFO-ANN Ensemble Hybrid Algorithm to Design Compact Fractal Antenna for Rectenna System, Dhaliwal, B.S. and Pattnaik, S.S. © 2016.

For the implementation of ANN ensemble of Fig. (**4.6**), the commonly used method of bagging (*i.e.*, bootstrap aggregating) is employed to select the diverse training data sets from the complete source data set. The bagging algorithm proposed by Yang and Browne [40] is used in the present work and is given below:

1. Initialize i = 1, iterate while $i \leq p$

2. Initialize j = 1, iterate while $j \leq q$

   *Let Rand Row = S * rand()*

   If Rand Row $\leq$ S

   Let S_i (j, All Columns) = D(Rand Row, All Columns);

   Else back to step 2;

3. Back to step 1.

4. Output the final training data sets generated by bagging method: S_1, S_2, S_3, ..., S_p.

where D represents the complete training data, S is size of D, q is the size of new training data, and p is the number of new training data items. This algorithm is used to generate ten different training subsets, which are then used for the training of the ensemble members. The size of the training subsets has varied from 20 to 25.

All ten individually trained ANN models are tested using the same test data set before using them for the ensemble design. The performance of the ensemble is also evaluated using the same test data set and is presented in Table **4.7**, which depicts a very low value of absolute error. The performance comparison of constituent ANN models and ANN ensemble on the basis of absolute error for the same test data set is given in Fig. (**4.7**), which illustrates that the ANN ensemble performance is more acceptable as compared to any individual ANN model because the error fluctuation is relatively less. Hence, the use of an ANN ensemble for this design is a better option as compared to individual ANN models. The trained ANN ensemble is helpful in the analysis of tapered CRF antenna, *i.e.*, it can be used to estimate the resonant frequency for the given dimensions of the base rectangle.

Table 4.7. ANN Ensemble Performance for Test Data.

S. No.	Inputs		Resonant Frequency f_r(GHz)	
	L (mm)	*W* (mm)	Desired Output	ANN Ensemble Output
1	31.058	46.326	2.4066	2.4217
2	31.92	47.929	2.3173	2.3656
3	32.564	45.316	2.3312	2.2470
4	33.18	41.634	2.3471	2.3051
5	34.161	46.324	2.2419	2.2191
6	34.794	39.054	2.3094	2.2409
7	35.447	47.325	2.1585	2.1692
8	36.358	42.881	2.1292	2.1737
9	36.476	45.444	2.1196	2.0126
10	37.702	43.791	2.0857	2.0081

Average Absolute Error = 0.0521

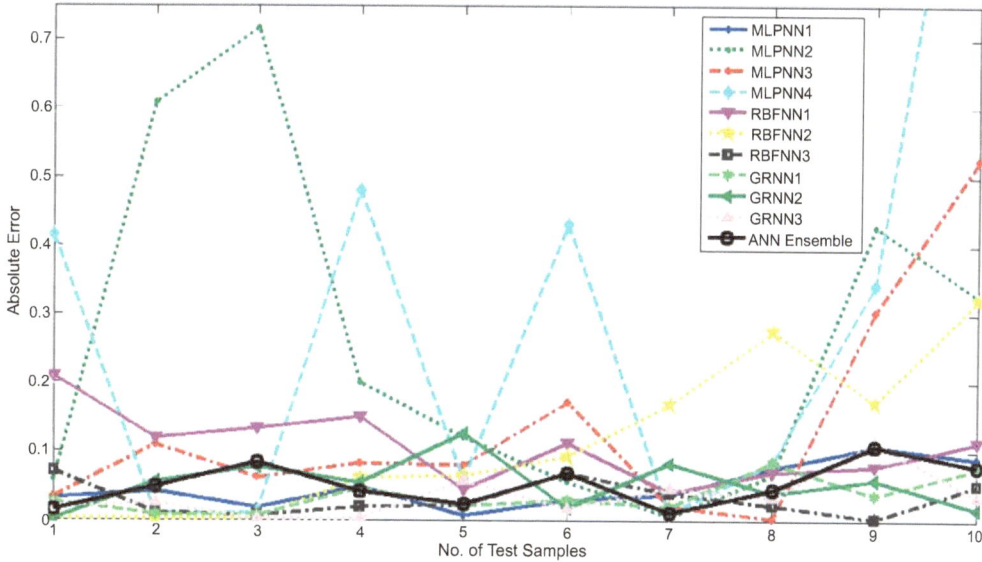

Fig. (4.7). Comparison of Absolute Error of Ensemble and Its Constituent ANN Models for Tapered CRF Antenna (Reprinted from the Springer Nature: Neural Computing and Applications, BFO-ANN Ensemble Hybrid Algorithm to Design Compact Fractal Antenna for Rectenna System, Dhaliwal, B.S. and Pattnaik, S.S. © 2016).

ANN Ensemble Model for CCF Antenna

An ANN ensemble model is explained in this section to estimate the resonant frequency of the first iteration CCF antenna described in Chapter 3 and shown in Fig. (**4.8**).

As discussed in Chapter 3, the f_r of the CCF antenna depends on the radius of the outer circle R. All other dimensions are calculated using R as explained in chapter 3. The ANN ensemble model having one input: the outer radius R and one output: the corresponding f_r, as shown in Fig. (**4.9**), is developed to analyze the CCF antenna. The ANN ensemble estimates the value of f_r for the given value of R. As shown in Fig. (**4.9**), nine diverse feed-forward ANN models which include three different MLPNNs trained with back-propagation algorithm, three different RBFNNs, and three different GRNNs are used to realize the ensemble [41]. The internal structures of individual ANN models are specified in Table **4.8**. All three MLPNNs have a different number of hidden layers, and the number of neurons is also different. The different spread constants of RBFNNs and GRNNs introduce the desired diversity in these models. Also, all the models are initialized with different random weights. The outputs of the individual ANN models are combined by a simple averaging method to generate the final ensemble output.

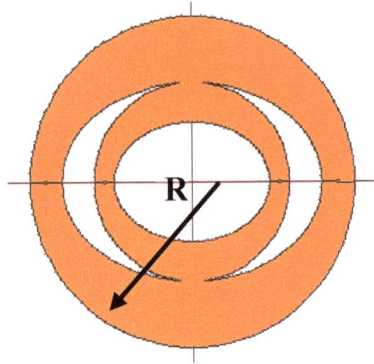

Fig. (4.8). First Iteration CCF Antenna [41].

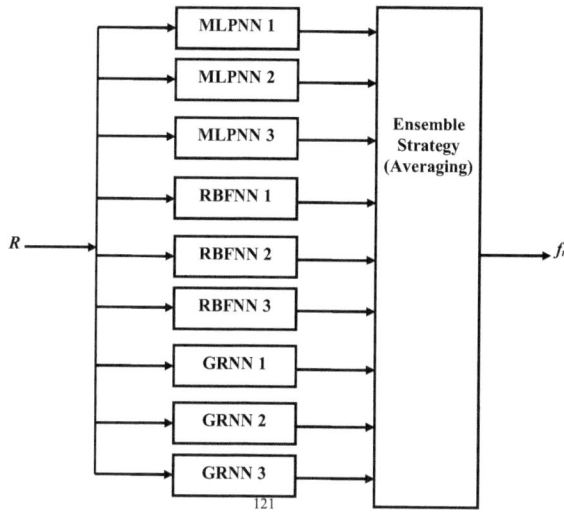

Fig. (4.9). ANN Ensemble for Resonant Frequency Estimation of CCF Antenna (Reprinted from the Springer Nature: Wireless Personal Communications, Development of PSO-ANN Ensemble Hybrid Algorithm and Its Application in Compact Crown Circular Fractal Patch Antenna Design, Dhaliwal, B.S. and Pattnaik, S.S. © 2017).

Table 4.8. Structures of the ANN Models used in the Ensemble for CCF Antenna.

S. No.	Network Name	Network Type	No. of Hidden Layers	No. of Hidden Neurons	Other Parameters
1	MLPNN 1	MLPNN	2	20, 20	Training Function = *Trainlm* Learning Rate = 0.2 Performance Goal = 10^{-6}
2	MLPNN 2	MLPNN	1	30	Training Function = *Trainlm* Learning Rate = 0.15 Performance Goal = 10^{-6}
3	MLPNN 3	MLPNN	3	10, 15, 10	Training Function = *Trainbfg* Learning Rate = 0.1 Performance Goal = 10^{-5}
4	RBFNN 1	RBFNN	1	21	Spread Constant = 0.25
5	RBFNN 2	RBFNN	1	21	Spread Constant = 0.55
6	RBFNN 3	RBFNN	1	21	Spread Constant = 1.0
7	GRNN 1	GRNN	2	21, 2	Spread Constant = 0.85
8	GRNN 2	GRNN	2	21, 2	Spread Constant = 0.7
9	GRNN 3	GRNN	2	21, 2	Spread Constant = 0.9

*(Reprinted from the Springer Nature: Wireless Personal Communications, Development of PSO-ANN Ensemble Hybrid Algorithm and Its Application in Compact Crown Circular Fractal Patch Antenna Design, Dhaliwal, B.S. and Pattnaik, S.S. © 2017).

IE3D software is used for generating the data set for the training and testing of different ANN models used in the ensemble. The 36 different values of R in the range 6 mm to 14 mm are randomly generated. The CCF antennas of these dimensions are simulated using the IE3D simulator and corresponding resonant frequencies are evaluated. The complete training and testing data set is shown in Table **4.9**. Out of these 36 samples, 9 samples marked with '*' are used for testing and the remaining 27 samples are used for training.

Table 4.9. Complete Data Set for ANN Ensemble Design.

S. No.	Radius of the Base Geometry R (mm)	Resonant Frequency f_r (GHz)	S. No.	Radius of the Base Geometry R (mm)	Resonant Frequency f_r (GHz)
1	6.0000	7.5573	19*	9.4672	4.8600
2	6.3000	7.3033	20	9.5362	4.8202
3*	6.5680	7.0012	21	9.8766	4.6584
4	6.7500	6.8996	22*	9.9000	4.6404
5	7.0000	6.5304	23	9.9442	4.6202
6	7.5000	6.1191	24	10.0000	4.6000
7	7.6071	6.0798	25*	10.4650	4.4233
8*	7.7926	5.9202	26	10.8470	4.3015
9	7.8810	5.9179	27	11.2460	4.1392
10	7.9257	5.8393	28*	11.6630	3.9899
11	7.9568	5.7898	29	12.0000	3.8393
12*	7.9728	5.7786	30	12.2840	3.7662
13	8.0000	5.7404	31	12.8560	3.6180
14	8.3355	5.5000	32	13.0000	3.5370
15	8.7653	5.2199	33*	13.3050	3.4612
16*	8.9561	5.1202	34	13.5000	3.4180
17	9.1406	5.0214	35	13.7320	3.3691
18	9.3971	4.9000	36	14.0000	3.3202

*used for testing "Reprinted from the Springer Nature: Wireless Personal Communications, Development of PSO-ANN Ensemble Hybrid Algorithm and Its Application in Compact Crown Circular Fractal Patch Antenna Design, Dhaliwal, B.S. and Pattnaik, S.S. © 2017.

The bagging (*i.e.*, bootstrap aggregating) algorithm discussed in previous section is employed to select the diverse training data sets from the complete training data set of (Table **4.9**). This method is used to generate 09 different training subsets which are then used for the training of the ensemble members. The size of the training subsets has been taken as 21. The trained ANN models are arranged to form ANN ensemble of (Fig. **4.9**). The same test data set is used to evaluate the performance of the ensemble and the constituent ANN models. The performance comparison is given in Fig. **(4.10)** which depicts a very low value of absolute error and the error fluctuation is relatively less for ANN ensemble so, it is better to use ensemble than individual ANN models. The trained ensemble model estimates the resonant frequency for the given value of base radius R instantaneously avoiding long simulation cycles. So, ANN ensemble method is a fast and accurate method of analyzing the CCF antenna behaviour.

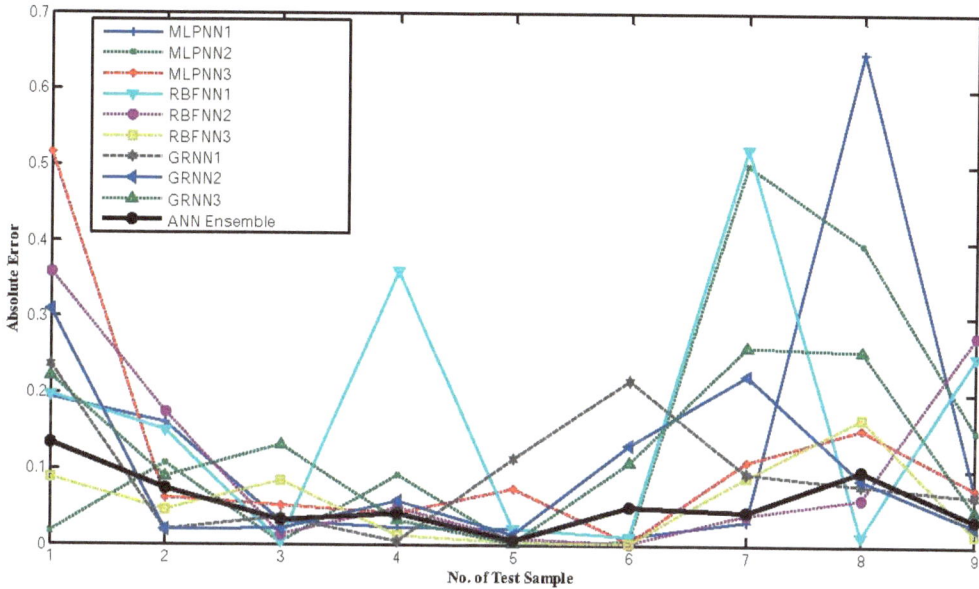

Fig. (4.10). Comparison of Absolute Error of Ensemble and Its Constituent ANN Models for CCF Antenna (Reprinted from the Springer Nature: Wireless Personal Communications, Development of PSO-ANN Ensemble Hybrid Algorithm and Its Application in Compact Crown Circular Fractal Patch Antenna Design, Dhaliwal, B.S. and Pattnaik, S.S. © 2017).

CONCLUSION

This chapter is devoted to the development of the ANN models, including ensemble models, for optimal parameter estimation of designed fractal antennas. The chapter starts with a brief discussion of the role of ANN in antennas. Firstly, models of ANNs are developed for the analysis of the SGMF antenna. Their performances are compared on the basis of *MAE* and C_R. It is found that the RBFNN model is more suitable for this type of application. The performance comparison of MLPNN, RBFNN and GRNN models for various parameter estimation of CRF antenna is also discussed. The designed models take feed location and iteration number as inputs and estimate the corresponding values of resonant frequency, S_{11} and gain of the antenna. The comparison on the basis of mean percentage error shows that the GRNN is the best choice for this application. The development of ANN ensemble models is a major contribution of this chapter. The first ANN ensemble model described in this chapter is used for estimating the resonant frequency of tapered CRF antenna for given values of antenna dimensions. Ten different ANN models are developed for the same input-output relationship using the diverse data set. The trained ANN models are then used to form an ensemble. It is observed that the ANN ensemble model has better performances than all of its constituent models. An ANN ensemble model for the analysis of CCF antenna is also discussed in this chapter. The ensemble model predicts the resonant frequency of the antenna for a given outer radius of the antenna. 09 diverse ANN models are developed to form the ensemble. The results of the ensemble are compared with those of constituent models to evaluate the benefits of the ensemble approach.

DISCLOSURE

Part of this article has previously been published in the following articles:

• B. S. Dhaliwal and S. S. Pattnaik, "Artificial neural network analysis of Sierpinski gasket fractal antenna: A low cost alternative to experimentation," *Adv. Artif. Neural Syst.*, vol. 2013, pp. 1–7, 2013.

• B. S. Dhaliwal and S. S. Pattnaik, "Performance evaluation of Artificial Neural Networks in microstrip fractal antenna parameter estimation," in *Proceedings of 2012 IEEE International Conference on Communication Systems (ICCS)*, 2012, pp. 135-139.

• B. S. Dhaliwal and S. S. Pattnaik, "BFO–ANN ensemble hybrid algorithm to design compact fractal antenna for rectenna system," *Neural Comput. Appl.*, vol. 28, no. S1, pp. 917–928, 2017.

• B. S. Dhaliwal and S. S. Pattnaik, "Development of PSO-ANN ensemble hybrid algorithm and its application in compact crown circular fractal patch antenna design," *Wirel. Pers. Commun.*, vol. 96, no. 1, pp. 135–152, 2017.

REFERENCES

[1]   R.K. Mishra, and A. Patnaik, "Designing rectangular patch antenna using the neurospectral method", *IEEE Trans. Antenn. Propag.*, vol. 51, no. 8, pp. 1914-1921, 2003. [http://dx.doi.org/10.1109/TAP.2003.814748]

[2]   N. Turker, F. Gunes, and T. Yildirim, "Artificial Neural Design of Microstrip Antennas", *Turk. J. Electr. Eng. Comput. Sci.*, vol. 14, no. 3, pp. 445-453, 2006.

[3]   S. Devi, D.C. Panda, S.S. Pattnaik, B. Khuntia, and D.K. Neog, "Initializing artificial neural networks by genetic algorithm to calculate the resonant frequency of single shorting post rectangular patch

antenna", *Proceedings of IEEE Antennas and Propagation Society International Symposium,* pp. 144-147, 2003.
[http://dx.doi.org/10.1109/APS.2003.1219810]

[4]   B. Khuntia, S.S. Pattnaik, D.C. Panda, D.K. Neog, S. Devi, and M. Dutta, "Genetic algorithm with artificial neural networks as its fitness function to design rectangular microstrip antenna on thick substrate", *Microw. Opt. Technol. Lett.,* vol. 44, no. 2, pp. 144-146, 2005.
[http://dx.doi.org/10.1002/mop.20570]

[5]   K.A. Kumar, R. Ashwath, D.S. Kumar, and R. Malmathanraj, "Optimization of multislotted rectangular microstrip patch antenna using ANN and bacterial foraging optimization", *Proceedings of 2010 Asia-Pacific International Symposium on Electromagnetic Compatibility,* pp. 449-452, 2010.
[http://dx.doi.org/10.1109/APEMC.2010.5475810]

[6]   V.S. Chintakindi, S.S. Pattnaik, O.P. Bajpai, and S. Devi, "PSO driven RBFNN for design of equilateral triangular microstrip patch antenna", *Indian J. Radio Space Phys.,* vol. 34, no. 4, pp. 233-237, 2009.

[7]   Z. Wang, S. Fang, Q. Wang, and H. Liu, "An ANN-based synthesis model for the single-feed circularly-polarized square microstrip antenna with truncated corners", *IEEE Trans. Antenn. Propag.,* vol. 60, no. 12, pp. 5989-5992, 2012.
[http://dx.doi.org/10.1109/TAP.2012.2214195]

[8]   K. Siakavara, "Artificial neural network based design of a three-layered microstrip circular ring antenna with specified multi-frequency operation", *Neural Comput. Appl.,* vol. 18, no. 1, pp. 57-64, 2009.
[http://dx.doi.org/10.1007/s00521-007-0153-3]

[9]   L.H. Manh, F. Grimaccia, M. Mussetta, and R.E. Zich, "Optimization of a dual ring antenna by means of artificial neural network", *Prog. Electromagn. Res. B Pier B,* vol. 58, pp. 59-69, 2014.
[http://dx.doi.org/10.2528/PIERB13112806]

[10]   I. Vilovic, N. Burum, and M. Brailo, "Microstrip antenna design using neural networks optimized by PSO", *Proceedings of International Conference on Applied Electromagnetics and Communications,* pp. 1-4, 2013.
[http://dx.doi.org/10.1109/ICECom.2013.6684759]

[11]   S.S. Gultekin, D. Uzer, and O. Dundar, "Calculation of circular microstrip antenna parameters with a single artificial neural network model", *Proceedings of Progress in Electromagnetics Research Symposium,* pp. 545-548, 2012.

[12]   T. Bose, and N. Gupta, "Design of an aperture-coupled microstrip antenna using a hybrid neural network", *IET Microw. Antennas Propag.,* vol. 6, no. 4, pp. 470-474, 2012,.
[http://dx.doi.org/10.1049/iet-map.2011.0363]

[13]   S. Devi, D.C. Panda, S.S. Pattnaik, B. Khuntia, and D.K. Neog, "Initializing artificial neural networks by genetic algorithm to calculate the resonant frequency of single shorting post rectangular patch antenna", *Proceedings of IEEE Antennas and Propagation Society International Symposium,* pp. 144-147, 2003.
[http://dx.doi.org/10.1109/APS.2003.1219810]

[14]   B. Khuntia, S.S. Pattnaik, D.C. Panda, D.K. Neog, S. Devi, and M. Dutta, "A simple and efficient approach to train artificial neural networks using a genetic algorithm to calculate the resonant frequency of an RMA on thick substrate", *Microw. Opt. Technol. Lett.,* vol. 41, no. 4, pp. 313-315, 2004.
[http://dx.doi.org/10.1002/mop.20126]

[15]   S.S. Pattnaik, B. Khuntia, D.C. Panda, D.K. Neog, S. Devi, and M. Dutta, "Application of a genetic algorithm in an artificial neural network to calculate the resonant frequency of a tunable single-shorting-post rectangular-patch antenna", *Int. J. RF Microw. Comput.-Aided Eng.,* vol. 15, no. 1, pp. 140-144, 2005.

[http://dx.doi.org/10.1002/mmce.20060]

[16]   Y. Tighilt, F. Bouttout, and A. Khellaf, "Modeling and design of printed antennas using neural networks", *Int. J. RF Microw. Comput.-Aided Eng.,* vol. 21, no. 2, pp. 228-233, 2011.
[http://dx.doi.org/10.1002/mmce.20509]

[17]   S. Can, K.Y. Kapusuz, and E. Aydin, "Calculation of resonant frequencies of a shorting pin-loaded ETMA with ANN", *Microw. Opt. Technol. Lett.,* vol. 56, no. 3, pp. 660-663, 2014.
[http://dx.doi.org/10.1002/mop.28157]

[18]   T. Yu-Bo, Z. Su-Ling, and L. Jing-Yi, "Modeling resonant frequency of microstrip antenna based on neural network ensemble", *Int. J. Numer. Model.,* vol. 24, no. 1, pp. 78-88, 2011.
[http://dx.doi.org/10.1002/jnm.761]

[19]   K. Hettak, and G.Y. Delisle, "Low profile cellular radio antenna for ISM applications", *Proceedings of IEEE Antennas and Propagation Society International Symposium,* pp. 443-446, 2004.

[20]   D.K. Neog, S.S. Pattnaik, D.C. Panda, S. Devi, B. Khuntia, and M. Dutta, "Design of a wideband microstrip antenna and the use of artificial neural networks in parameter calculation", *IEEE Antennas Propag. Mag.,* vol. 47, no. 3, pp. 60-65, 2005.
[http://dx.doi.org/10.1109/MAP.2005.1532541]

[21]   Y. Kim, S. Keely, J. Ghosh, and H. Ling, "Application of artificial neural networks to broadband antenna design based on a parametric frequency model", *IEEE Trans. Antenn. Propag.,* vol. 55, no. 3, pp. 669-674, 2007.
[http://dx.doi.org/10.1109/TAP.2007.891564]

[22]   D.C. Panda, S.S. Pattnaik, S. Devi, and R.K. Mishra, "Application of FIR-neural network on finite difference time domain technique to calculate input impedance of microstrip patch antenna", *Int. J. RF Microw. Comput.-Aided Eng.,* vol. 20, no. 2, pp. 158-162, 2010.
[http://dx.doi.org/10.1002/mmce.20417]

[23]   I. Vilovic, and N. Burum, "Design and feed position estimation for circular microstrip antenna based on neural network model", *Proceedings of 6th European Conference on Antennas and Propagation (EUCAP),* pp. 3614-3617, 2012.
[http://dx.doi.org/10.1109/EuCAP.2012.6206281]

[24]   M. Aneesh, J.A. Ansari, A. Singh, Kamakshi, and S.S. Sayeed, "Kamakshi, and S. S. Sayeed, Analysis of microstrip line feed slot loaded patch antenna using artificial neural network", *Prog. Electromagn. Res. B Pier B,* vol. 58, pp. 35-46, 2014.
[http://dx.doi.org/10.2528/PIERB13111105]

[25]   A. Patnaik, B. Choudhury, P. Pradhan, R.K. Mishra, and C. Christodoulou, "An ANN application for fault finding in antenna arrays", *IEEE Trans. Antenn. Propag.,* vol. 55, no. 3, pp. 775-777, 2007.
[http://dx.doi.org/10.1109/TAP.2007.891557]

[26]   P.N. Bashly, and E.L. Popovskii, "Optimization of antenna arrays based on neurocomputers", *Autom. Control Comput. Sci.,* vol. 41, no. 3, pp. 141-147, 2007.
[http://dx.doi.org/10.3103/S0146411607030042]

[27]   K.C. Lee, J.Y. Jhang, and T.N. Lin, "An automatically converging scheme based on the neural network and its application in antennas", *IEEE Trans. Antenn. Propag.,* vol. 57, no. 4, pp. 1270-1274, 2009.
[http://dx.doi.org/10.1109/TAP.2009.2015856]

[28]   D. Vakula, and N.V.S.N. Sarma, "Fault diagnosis of planar antenna arrays using neural networks", *Prog. Electromagn. Res. M Pier M,* vol. 6, pp. 35-46, 2009.
[http://dx.doi.org/10.2528/PIERM09011204]

[29]   Z.D. Zaharis, C. Skeberis, T.D. Xenos, P.I. Lazaridis, and J. Cosmas, "Design of a novel antenna array beamformer using neural networks trained by modified adaptive dispersion invasive weed optimization based data", *IEEE Trans. Broadcast,* vol. 59, no. 3, pp. 455-460, 2013.

[http://dx.doi.org/10.1109/TBC.2013.2244793]

[30]   S. Mishra, R.N. Yadav, and R.P. Singh, "Directivity estimations for short dipole antenna arrays using radial basis function neural networks", *IEEE Antennas Wirel. Propag. Lett.,* vol. 14, pp. 1219-1222, 2015.
[http://dx.doi.org/10.1109/LAWP.2015.2399453]

[31]   Qi-Jun Zhang, K.C. Gupta, and V.K. Devabhaktuni, "Artificial neural networks for rf and microwave design-from theory to practice", *IEEE Trans. Microw. Theory Tech.,* vol. 51, no. 4, pp. 1339-1350, 2003.
[http://dx.doi.org/10.1109/TMTT.2003.809179]

[32]   C. Puente-Baliarda, J. Romeu, R. Pous, and A. Cardama, "On the behavior of the Sierpinski multiband fractal antenna", *IEEE Trans. Antenn. Propag.,* vol. 46, no. 4, pp. 517-524, 1998.
[http://dx.doi.org/10.1109/8.664115]

[33]   C.T.P. Song, P.S. Hall, and H. Ghafouri-Shiraz, "Perturbed Sierpinski multiband fractal antenna with improved feeding technique", *IEEE Trans. Antenn. Propag.,* vol. 51, no. 5, pp. 1011-1017, 2003.
[http://dx.doi.org/10.1109/TAP.2003.811522]

[34]   R. Mishra, R. Ghatak, and D. Poddar, "Design formula for Sierpinski gasket pre-fractal planar-monopole antennas", *IEEE Antennas Propag. Mag.,* vol. 50, no. 3, pp. 104-107, 2008.
[http://dx.doi.org/10.1109/MAP.2008.4563575]

[35]   B.S. Dhaliwal, and S.S. Pattnaik, "Artificial neural network analysis of Sierpinski gasket fractal antenna: A low cost alternative to experimentation", *Adv. Artif. Neural Syst.,* vol. 2013, pp. 1-7, 2013.
[http://dx.doi.org/10.1155/2013/560969]

[36]   B.S. Dhaliwal, and S.S. Pattnaik, "Performance evaluation of Artificial Neural Networks in microstrip fractal antenna parameter estimation", *Proceedings of 2012 IEEE International Conference on Communication Systems (ICCS),* pp. 135-139, 2012.
[http://dx.doi.org/10.1109/ICCS.2012.6406124]

[37]   M.M. Islam, Xin Yao, and K. Murase, "A constructive algorithm for training cooperative neural network ensembles", *IEEE Trans. Neural Netw.,* vol. 14, no. 4, pp. 820-834, 2003.
[http://dx.doi.org/10.1109/TNN.2003.813832] [PMID: 18238062]

[38]   B.S. Dhaliwal, and S.S. Pattnaik, "BFO–ANN ensemble hybrid algorithm to design compact fractal antenna for rectenna system", *Neural Comput. Appl.,* vol. 28, no. S1, pp. 917-928, 2017.
[http://dx.doi.org/10.1007/s00521-016-2402-9]

[39]   N. García-Pedrajas, "Constructing ensembles of classifiers by means of weighted instance selection", *IEEE Trans. Neural Netw.,* vol. 20, no. 2, pp. 258-277, 2009.
[http://dx.doi.org/10.1109/TNN.2008.2005496] [PMID: 19179252]

[40]   S. Yang, and A. Browne, "Neural network ensembles: combining multiple models for enhanced performance using a multistage approach", *Expert Syst.,* vol. 21, no. 5, pp. 279-288, 2004.
[http://dx.doi.org/10.1111/j.1468-0394.2004.00285.x]

[41]   B.S. Dhaliwal, and S.S. Pattnaik, "Development of PSO-ANN ensemble hybrid algorithm and its application in compact crown circular fractal patch antenna design", *Wirel. Pers. Commun.,* vol. 96, no. 1, pp. 135-152, 2017.
[http://dx.doi.org/10.1007/s11277-017-4157-8]

Development of Hybrid Bio-inspired Computing Algorithms for Design of Fractal Antennas

Abstract: One of the novel contributions of this book is the development of hybrid bio-inspired computing algorithms for the design of fractal antennas. This work is presented in this chapter. The hybrid algorithms are developed to design the proposed fractal antennas for desired frequencies. The performance comparison of bio-inspired computing algorithms for the design of a multiband Sierpinski Gasket fractal antenna is also explained. The development of various hybrid algorithms like the GA-ANN hybrid Algorithm, BFO-ANN ensemble hybrid Algorithm, and PSO-ANN Ensemble hybrid Algorithm is explained. The use of ANN models as objective functions of optimization algorithms is discussed in this chapter. This chapter also deals with the experimental testing and validation of the developed fractal antennas. The photographs of the fabricated antennas and the experimental results are included. The comparison of the simulated results and experimental results is discussed. The suitability of the designed antennas for different applications is also highlighted in this chapter.

Keywords: Bacterial foraging optimization, Crown fractal antenna, Fractal antenna, Genetic algorithms, Miniaturized antenna, Particle swarm optimization, Sierpinski gasket.

INTRODUCTION

One of the challenging points in designing an antenna for a given frequency is to determine the required accurate dimensions of the antenna [1]. In fractal antennas, the radiating patches use complex geometries with many parameters, so the method used to determine the dimensions must consider all the geometry parameters. Also, the closed-form expressions do not exist, and the development of analytical methods is extremely difficult for complex fractal geometries of antennas. The numerical modeling methods are computationally expensive, and empirical modeling solutions have limited range and accuracy. In such cases, the use of bio-inspired optimization algorithms is very suitable [2, 3]. In the last decade, there has been exponential growth in the use of artificial intelligence techniques like ANN, fuzzy logic systems and bio-inspired optimization echniques in the designing of antennas [4 - 8]. The two most popular bio-inspired optimization algorithms: GA and PSO have been applied by a few researchers for

Balwinder S. Dhaliwal, Suman Pattnaik & Shyam Sundar Pattnaik the optimization of fractal antennas in recent years [9 - 13]. But, the application of another popular bio-inspired optimization algorithm, the BFO algorithm, in the field of fractal antennas is not investigated yet.

Literature shows that the initial work in the domain of optimization of fractal wire antennas is proposed by Werner *et al.* [11]. Pantoja *et al.* [14] proposed the optimization of fractal shapes such as the Delta, Koch, and Sierpinski-types using a multi-objective GA. A procedure for designing Sierpinski gasket and Koch monopole fractal antennas for user-defined frequencies using ANN and the PSO technique is presented by Anuradha *et al.* [10]. Pantoja *et al.* [12] applied a multi-objective GA algorithm to the design of wire fractal antennas optimizing their bandwidth and efficiency while reducing their resonant frequencies. Azaro *et al.* [15] described the design of a Koch-like fractal miniaturized monopole antenna for ISM-band application by optimizing the fractal geometry and the segment widths through a PSO algorithm. A PSO-based approach for the design and optimization of a triple-band fractal-eroded antenna has been described [16]. The synthesis of a miniaturized three-band planar antenna working in GSM and Wi-Fi frequency bands is described [17]. Lizzi and Massa [18] proposed a dual-band fractal monopole antenna based on a perturbed planar Sierpinski fractal shape suitable for LTE standards by means of PSO. An H-shaped fractal antenna is optimized by Weng and Hung [19] using the PSO algorithm for 2.45 GHz and 5.5 GHz WLAN applications. In bio-inspired optimization algorithms, a suitable objective function relating the variables to be optimized with the design variables is required. As the mathematical expressions are not available for fractal antennas, the use of the ANN model as an objective function is very appropriate in such cases [10, 20]. The use of ANN as an objective function of a GA for designing a rectangular microstrip antenna is described by Khuntia *et al.* [20]. The design of multiband fractal antennas using the PSO algorithm employing the ANN as an objective function is presented by Anuradha *et al.* [10].

The comparative analysis of different approaches available to solve a specific problem is important to find the best suitable algorithm for that application, and in the past, many researchers have compared the performance of different analytical and optimization algorithms for a variety of antennas. A comparison of different models to find the resonant frequencies of a rectangular microstrip antenna is proposed by Dearnley and Barel [1]. An analytical antenna model for chiral scatterers is compared with numerical and experimental data by Tretyakov *et al.* [21]. The application of GA and PSO for the phased array synthesis is compared by Boeringer and Werner [22]. The performance of optimized signal processing algorithms is analyzed for smart antenna systems by Gondal and Anees [23]. Pérez and Basterrechea [24] presented the performance of different heuristic optimization methods for radiation pattern estimation. The performance of BFO and PSO algorithms for adaptive antenna array processing is compared by Datta and Misra [25]. Panduro *et al.* [26] compared the performance of three different evolution methods for the design of scannable circular antenna arrays. However, the comparative analysis of optimization algorithms for fractal antennas has not been explored yet.

In this chapter, the applications of bio-inspired computing techniques to design fractal antennas are described. The performance comparison of different methods on the basis of different metrics is also discussed. The development of hybrids of optimization algorithms and ANN is also presented. The standard method of validation of the performance of antennas designed using simulation software is by fabricating the prototypes of antennas and then experimentally measuring the various parameters. This chapter also presents the measured results of the prototypes of proposed antennas and compares them with the simulation results. The potential applications of the fabricated antennas are also discussed.

DESIGN OF SGMF ANTENNA USING BIO-INSPIRED COMPUTING TECHNIQUES

As discussed in Chapter 4, the latest updated expression to determine the frequency of SGMF antenna for a particular side-length s is proposed by Mishra *et al.* [27]. This formula, as shown in equation (5.1), predicts the value of f_r of the SGMF antenna for the given values of s, ε_r, h and n.

$$f_r = \begin{cases} (0.15345 + 0.34\rho x)\frac{c}{H_e}\,(\xi^{-1})^n & for \; n = 0 \\ 0.26\frac{c}{H_e}\delta^n & for \; n > 0 \end{cases} \tag{5.1}$$

In equation (5.1), H_e denotes the effective height of the SGMF antenna, and it is calculated as $H_e = \frac{\sqrt{3}s_e}{2}$, s_e is the effective side-length of the SGMF antenna, and it is given by $s_e = s + \frac{h}{\sqrt{\varepsilon_r}}$, c is speed of light, δ is scale ratio, and its value is 2 for antenna under consideration, ξ is defined as equal to $1/\delta$, ρ is equal to ξ - 0.230735, and x is defined as 0 for $n = 0$.

By putting the above values of H_e, s_e, c, δ, ξ, ρ, and x in equation (5.1), the simplified expression for calculating f_r of SGMF antenna is proposed in this thesis as equation (5.2):

$$f_r = \begin{cases} 53.157\,\frac{\sqrt{\varepsilon_r}}{h+s\sqrt{\varepsilon_r}}\,10^9 & for \; n = 0 \\ 90.067\,\frac{2^n\sqrt{\varepsilon_r}}{h+s\sqrt{\varepsilon_r}}\,10^9 & for \; n > 0 \end{cases} \tag{5.2}$$

For the designing of SGMF antenna, it is required to calculate the side-length for the given resonant frequency and substrate parameters. Mishra *et al.* [27] also proposed an expression for calculating side-length *s* by rearranging the equation (5.1) as given below in equation (5.3):

$$s \cong \begin{cases} \frac{1}{\sqrt{3}}(0.3069+0.68\rho x)\frac{c}{fr}(\xi^{-1})^n - \frac{h}{\sqrt{\varepsilon_r}} & for\ n=0 \\ \frac{0.52}{\sqrt{3}}\frac{c}{fr}(\delta)^n - \frac{h}{\sqrt{\varepsilon_r}} & for\ n>0 \end{cases} \tag{5.3}$$

The simplified version of equation (5.3) is also obtained by putting the values of various parameters defined above, and the proposed expression is expressed in equation (5.4):

$$s \cong \begin{cases} 0.177\frac{c}{fr} - \frac{h}{\sqrt{\varepsilon_r}} & for\ n=0 \\ 0.301\frac{c}{fr}(2)^n - \frac{h}{\sqrt{\varepsilon_r}} & for\ n>0 \end{cases} \tag{5.4}$$

In this section, an approach to design an SGMF antenna using bio-inspired optimization algorithms is presented. The simplified expression of equation (5.2) has been used as the fitness function of the optimization algorithms. The side-length *s* is taken as the design variable for the particular values of ε_r, *h* and *n*. This approach is implemented using three popular bio-inspired optimization algorithms: GA, PSO, and BFO. A large number of modifications in these algorithms have been proposed in recent years. However, in this work, only the standard forms of these optimization algorithms are considered in order to demonstrate the capabilities of the design procedure. Two different SGMF antennas are designed using these optimization algorithms [28].

Parameters of the Proposed Optimization Models for SGMF Antenna

The details of the parameters of the models developed for the SGMF antennas are described below:

The parameter values of the GA model are: Population size = 30; selection strategy = stochastic uniform; probability of crossover = 0.80; mutation function = Gaussian; percentage of parents' population being replaced by children = 93.3%; and number of maximum generations = 200. The PSO algorithm starts by initializing a group of 30 particles, with random positions constrained between 0 and 100. A set of random velocities is also initialized, with values between 0 and 1. A set of 100 iterations is used to find the optimum value in each case. The parameters used for BFO model are: number of bacteria used for searching the total region = 30; swimming length = 4; number of iterations taken in a chemotaxis loop = 50; number of reproductions = 4; elimination and dispersal number = 5; and probability of elimination and dispersal= 0.25.

Results of the Proposed Optimization Models for SGMF Antenna

The results of the optimization models for the design of two different SGMF antennas are presented below. The error graphs for the developed optimization algorithms are shown in Fig. (**5.1**), which depict that all the algorithms converge to almost zero value of error.

The results of first SGMF antenna for the values of ε_r = 2.5 and h = 1.588 mm are shown in Tables **5.1.** and **5.2** shows the results of the second SGMF antenna for the values of ε_r = 2.4 and h = 1.59 mm. A comparison of results is prepared by determining the Absolute Percentage Error (*APE*) between experimentally measured values and those calculated by various models. Tables **5.1.** and **5.2** show that the percentage of error is very less for all models. The time taken by each model (run on a computer with 2 GHz Intel Core Duo processor having RAM of 4GB) for calculating the results is also presented for comparison purposes, and it shows that the PSO takes minimum time.

Table 5.1. *: Comparison of Results of First SGMF Antenna for ε_r = 2.5 and h = 1.588 mm.

S No.	f_r (GHZ)	n	Experi-mental value from [29]	Theoretical value from [27]		GA Results			PSO Results			BFO Results		
			s (mm)	s (mm)	*APE*	s (mm)	*APE*	Time Taken (msec)	s (mm)	*APE*	Time Taken (msec)	s (mm)	*APE*	Time Taken (msec)
1	0.52	0	102.77	101.22	1.51	101.23	1.50	494	101.23	1.50	158	101.42	1.31	5335
2	1.74	1	102.77	102.52	0.24	102.52	0.24	539	102.54	0.22	175	102.73	0.04	5421
3	3.51	2	102.77	101.64	1.10	101.64	1.10	494	101.64	1.10	173	101.72	1.02	5412
4	6.95	3	102.77	102.67	0.10	102.67	0.10	492	102.67	0.10	153	102.77	0.00	5496
5	13.89	4	102.77	102.74	0.03	102.76	0.01	501	102.76	0.01	183	102.78	0.01	5512

*Reprinted from the Springer Nature: Neural Computing and Applications, Performance Comparison of Bio-Inspired Optimization Algorithms for Sierpinski Gasket Fractal Antenna Design, Dhaliwal, B.S. and Pattnaik, S.S. © 2016.

The comparative analysis of the proposed models is presented on the basis of three different parameters, *i.e.*, (i) Mean Absolute Percentage Error (*MAPE*), which measures the performance of the model. Smaller the value of *MAPE* of a model better will be its performance; (ii) the average time taken by the models to evaluate the results, the less time it takes means faster is the algorithm and (iii) C_R, which is a measure of strength of the linear relationship developed by a particular model. The values of C_R close to 1.0 indicate a good performance of the model. The obtained results are compared with the published experimental results, which show that the *MAPE* is less than 1% for all models. It has been observed that all three algorithms outperform the theoretical method in accuracy because bio-inspired optimization algorithms search for the in-between solutions, which are not possible in theoretical methods. Due to the fast evolution properties of the models, the simulation time required is very less, and it is least in the case of PSO because the PSO evaluates only two steps for the calculation of new populations as compared to four in the case of GA and five for BFO. However, the BFO is a more compact search method as the step taken by the bacteria is less than the step-size of the particles in PSO and step-size of GA. So, the number of search points is more in BFO as compared to the other two optimization algorithms. As a result, it yields minimum error but takes more computational time. Among the different models compared, the PSO model is the best suitable model for this type of antenna design as it takes the least time to meet the required accuracy.

Table 5.2. *: Comparison of Results of Second SGMF Antenna for εr = 2.4 and h = 1.59 mm.

S No.	f_r (GHz)	n	Experimental Value from [29]	Theoretical Value from [27]		GA Results			PSO Results			BFO Results		
			s (mm)	s (mm)	APE	s (mm)	APE	Time Taken (msec)	s (mm)	APE	Time Taken (msec)	s (mm)	APE	Time Taken (msec)
1	0.880	0	60	59.38	1.03	59.39	1.02	414	59.38	1.03	180	59.58	0.70	4968
2	2.952	1	60	59.99	0.02	60.00	0.00	409	59.99	0.02	181	60.00	0.00	5028
3	5.982	2	60	59.20	1.33	59.20	1.33	408	59.20	1.33	197	59.34	1.10	4931
4	12.119	3	60	58.43	2.62	58.43	2.62	423	58.43	2.62	170	58.58	2.37	5031
5	23.610	4	60	60.01	0.02	60.01	0.02	420	59.99	0.02	170	60.01	0.02	4979

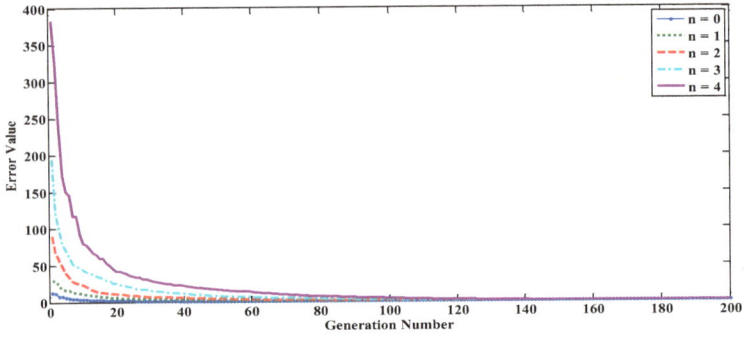

(a) Error Graph of GA for First SGMF Antenna

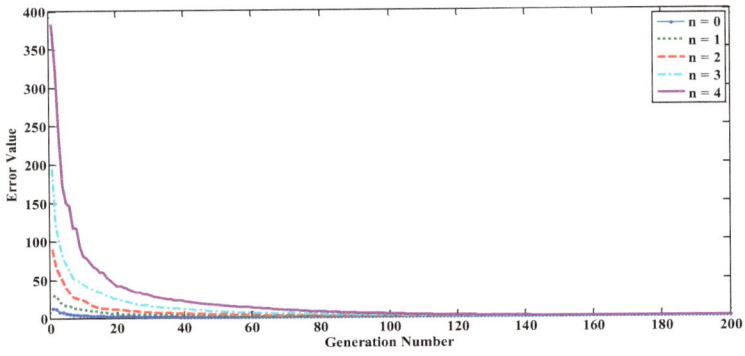

(b) Error Graph of GA for Second SGMF Antenna

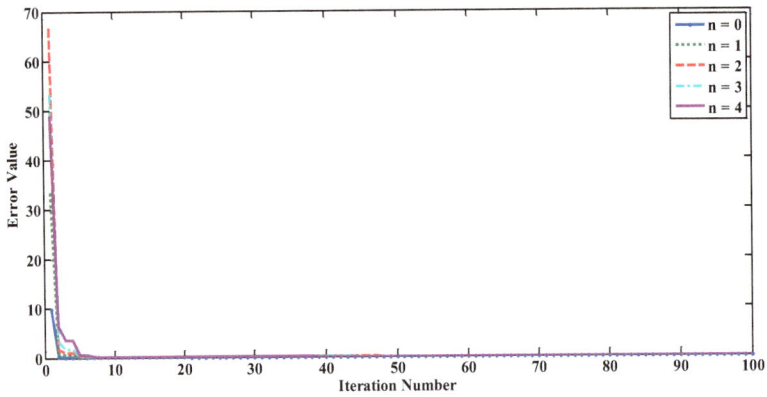

(c) Error Graph of PSO Algorithm for First SGMF Antenna

(Fig. 5.1) contd.....

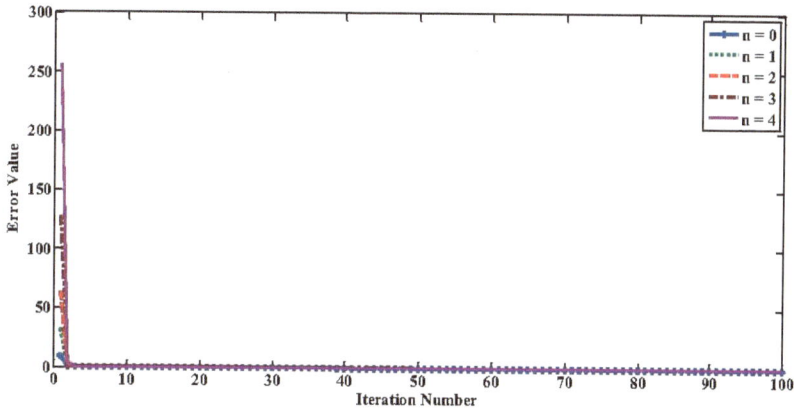

(d) Error Graph of PSO Algorithm for Second SGMF Antenna

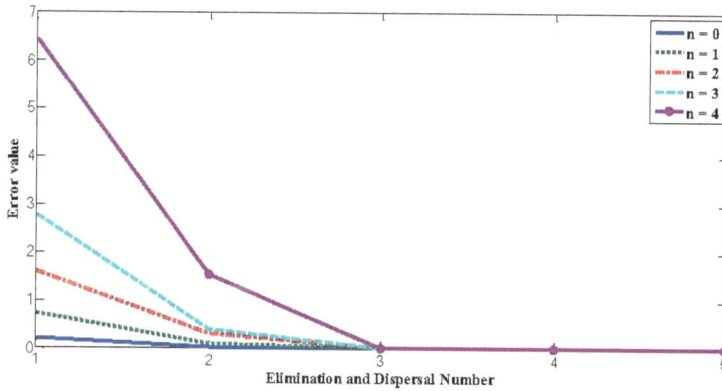

(e) Error Graph of BFO Algorithm for First SGMF Antenna

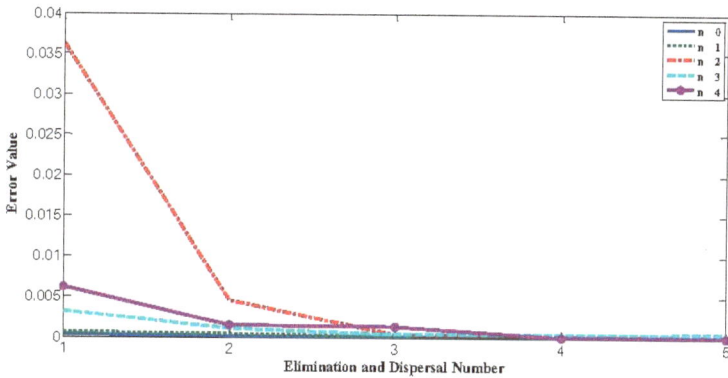

(f) Error Graph of BFO Algorithm for Second SGMF Antenna

Fig. (5.1). Error Graphs of Proposed Optimization Models for SGMF Antennas (Reprinted from the Springer Nature: Neural Computing and Applications, Performance Comparison of Bio-Inspired Optimization Algorithms for Sierpinski Gasket Fractal Antenna Design, Dhaliwal, B.S. and Pattnaik, S.S. © 2016).

GA-ANN HYBRID MODEL FOR SGMF ANTENNA DESIGN

This section describes the design of SGMF antenna using the GA-ANN hybrid algorithm. The ANN models for estimating the f_r of SGMF antenna for different values of s, n, ε_r, and h of the antenna are described in Chapter 4, and it has been observed that the MLPNN and RBFNN models have better performance than other models. As presented in Chapter 4, the MLPNN and RBFNN models have better performance than the other approaches, so if these ANN models are taken as the objective functions of the bio-inspired algorithms, then improved results are expected. This is explored in this section using the GA-ANN hybrid algorithm in which the MLPNN model is employed as the objective function of GA. The value of s is taken as the design variable for the particular values of ε_r, h, and n. The objective function used for this hybrid algorithm is presented below in equation (5.5):

$$Error = abs(f_d - f_r) \qquad (5.5)$$

where f_d is the desired resonant frequency for which the SGMF antenna is to be designed, f_r is the value of resonant frequency calculated using the MLPNN model for the given value of s. The absolute value of error given by the equation (5.5) is minimized using the GA and the value of s for the minimum value of error is the required optimal value of the side-length of SGMF antenna.

The above GA-ANN hybrid algorithm is used to design two different SGMF antennas. The first SGMF antenna has the values of $\varepsilon_r = 2.5$ and $h = 1.588$ mm and the second antenna has the values of $\varepsilon_r = 2.4$ and $h = 1.59$ mm. The parameters of the GA are: population size = 30; selection strategy = stochastic uniform; probability of crossover = 0.80; mutation function = Gaussian; percentage of parents' population being replaced by children = 93.3%; and number of maximum generations = 200. The ANN model is the MLPNN model for SGMF antenna developed in chapter 4. The results of the proposed hybrid algorithm are given in Table **5.3**, which also shows the experimental values and the results calculated by various other models.

Table 5.3. Comparison of Results of Different Models for SGMF Antenna.

S No.	f_r (GHz)	n	ε_r	h (mm)	Experi-mental Value [27, 29]	Theoretical Value [27]	GA Results [28]	PSO Results [28]	BFO Results [28]	GA-ANN Results
					s (mm)	s (mm)	s (mm)	s (mm)	s (mm)	s (mm)
1	0.520	0	2.5	1.588	102.77	101.22	101.23	101.23	101.42	101.90

(Table 5.3) cont.....

S No.	f_r (GHz)	n	ε_r	h (mm)	Experi-mental Value [27, 29]	Theoretical Value [27]	GA Results [28]	PSO Results [28]	BFO Results [28]	GA-ANN Results
					s (mm)	s (mm)	s (mm)	s (mm)	s (mm)	s (mm)
2	0.880	0	2.4	1.590	60.00	59.38	59.39	59.38	59.58	59.99
3	1.740	1	2.5	1.588	102.77	102.52	102.52	102.54	102.73	102.68
4	2.952	1	2.4	1.590	60.00	59.99	60.00	59.99	60.00	60.01
5	3.510	2	2.5	1.588	102.77	101.64	101.64	101.64	101.72	101.83
6	5.982	2	2.4	1.590	60.00	59.20	59.20	59.20	59.34	59.82
7	6.950	3	2.5	1.588	102.77	102.67	102.67	102.67	102.77	102.74
8	12.119	3	2.4	1.590	60.00	58.43	58.43	58.43	58.58	58.86
9	13.890	4	2.5	1.5880	102.77	102.74	102.76	102.76	102.78	102.75
10	23.610	4	2.4	1.590	60.00	60.01	60.01	59.99	60.01	60.01
Total Absolute Error						**6.0700**	**6.02**	**6.02**	**4.96**	**3.30**

The performance comparison on the basis of total absolute error shown in Table **5.3** depicts that the GA-ANN hybrid algorithm has the least value of total absolute error. So, the performance of the GA-ANN hybrid algorithm is better than the latest theoretical expression proposed by Mishra *et al.* [27] and the bio-inspired optimization algorithms described in the previous section. Therefore, the SGMF antenna can be designed for desired frequencies using the proposed GA-ANN hybrid algorithm.

BFO-ANN ENSEMBLE HYBRID ALGORITHM TO DESIGN TAPERED CRF ANTENNA FOR ISM BAND APPLICATIONS

A tapered CRF antenna is described in chapter 3, and it has two design variables, *i.e.*, the resonant frequency of this tapered CRF antenna depends on the values of L and W of the base rectangular shape. An ANN ensemble model for estimating the f_r of this tapered CRF antenna is developed in chapter 4. The design of this tapered CRF antenna for ISM band applications using a BFO-ANN ensemble hybrid model is explained in this section. The ANN ensemble model developed in chapter 4 is employed as an objective function of a BFO algorithm to calculate the optimal dimensions for resonant frequency of 2.45 GHz. The error function given in equation (5.6) is used as the fitness function of the BFO algorithm. The f_r is the instantaneous resonant frequency which is calculated using the designed ANN ensemble. The L and W of the tapered CRF antenna are taken as design variables [30].

$$Error = \sqrt[2]{(2.45 - f_r)^2} \qquad (5.6)$$

The initial parameters selected for the BFO algorithm are:

- *Dimension of the search space (p) = 2*
- *Total number of bacteria in the population (S) = 25*
- *Number of chemotactic steps (N_c) = 50*
- *Swimming length (N_s) = 4*
- *Number of reproduction steps (N_re) = 4*
- *Number of elimination–dispersal events (N_ed) = 3, and*
- *Elimination-dispersal probability (P_ed) = 0.35*

The flow chart of the proposed hybrid algorithm is shown in Fig. (**5.2**). The proposed hybrid algorithm will converge to its minimum error value (ideally 0) for optimal dimensions of the tapered CRF antenna resonating at 2.45 GHz. The optimal values of the dimensions L and W of the tapered CRF antenna provided by the algorithm are 32.0024 mm and 38.9978 mm, respectively which are rounded off to the nearest integers as 32 mm and 39 mm, respectively. Therefore, the tapered CRF antenna having the $L = 32$ mm and $W = 39$ mm should have f_r of 2.45 GHz. To verify this, the tapered CRF antenna of these optimal dimensions, as shown in Fig. (**5.3**), is simulated using the IE3D software and the S_{11} result obtained is shown in Fig. (**5.4**). The S_{11} plot shows that the optimal antenna resonates at a frequency of 2.4531 GHz and has a bandwidth of 152 MHz. The feed point of the antenna has been finalized as (6.95, 8.95) by taking centre of the antenna at (0, 0) using the trial-and-error method. The coding steps for the above described hybrid algorithm are presented in Annexure 5.1

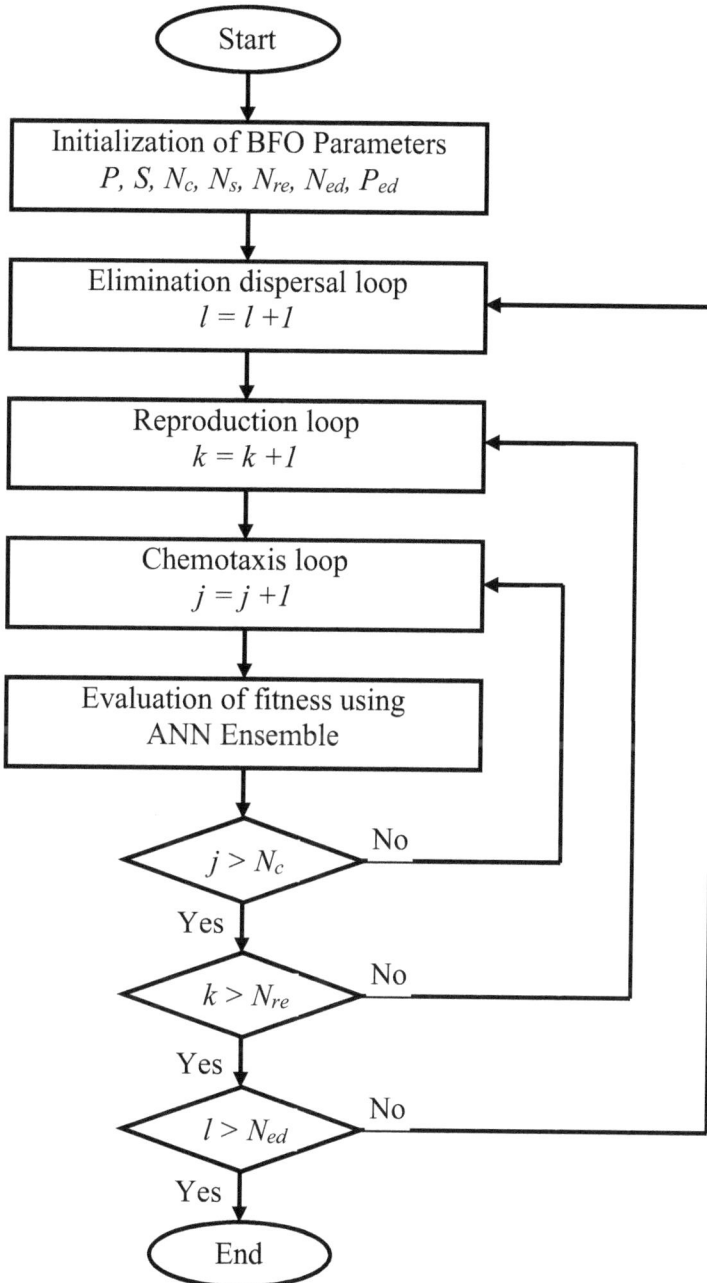

Fig. (5.2). BFO-ANN Ensemble Hybrid Model for Tapered CRF Antenna (Reprinted from the Springer Nature: Neural Computing and Applications, BFO-ANN Ensemble Hybrid Algorithm to Design Compact Fractal Antenna for Rectenna System, Dhaliwal, B.S. and Pattnaik, S.S. © 2016).

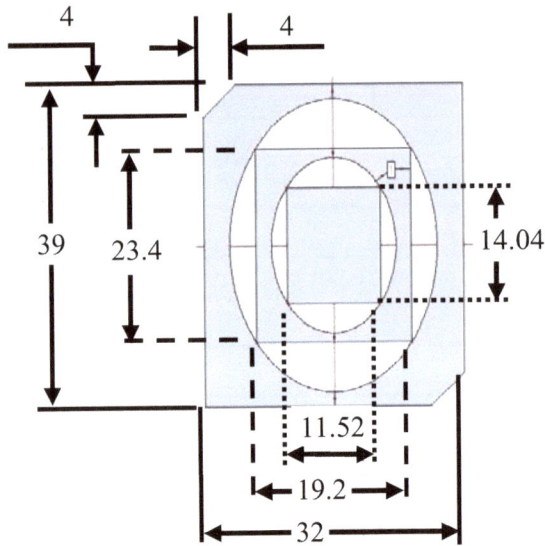

Fig. (5.3). Optimal Dimensions (in mm) of Tapered CRF Antenna for 2.45 GHz (Reprinted from the Springer Nature: Neural Computing and Applications, BFO-ANN Ensemble Hybrid Algorithm to Design Compact Fractal Antenna for Rectenna System, Dhaliwal, B.S. and Pattnaik, S.S. © 2016).

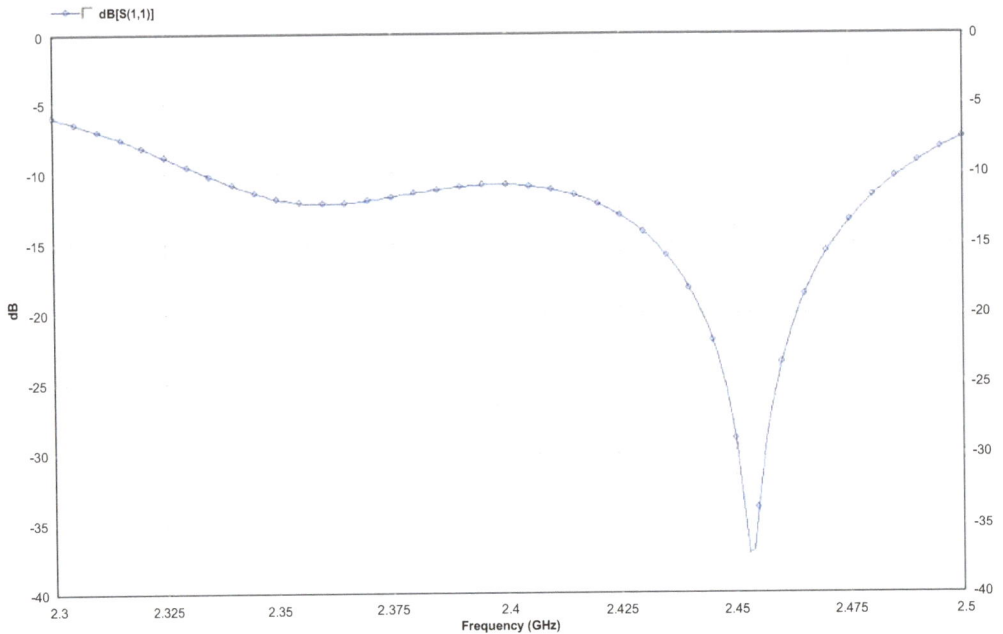

Fig. (5.4). S_{11} Plot of Optimal Tapered CRF Antenna (Reprinted from the Springer Nature: Neural Computing and Applications, BFO-ANN Ensemble Hybrid Algorithm to Design Compact Fractal Antenna for Rectenna System, Dhaliwal, B.S. and Pattnaik, S.S. © 2016).

As shown in Fig. (**5.4**), the optimal tapered CRF antenna resonates at 2.4531 GHz with the overall dimensions of $L = 32$ mm and $W = 39$ mm, resulting in an area of 1248 mm², however, the simple rectangular antenna for the same frequency requires the dimensions of 39.3 mm and 48.4 mm with an area of 1902.12 mm². So, the proposed optimal antenna has an area of 65.61% of the required area at 2.45 GHz resulting in a size reduction of 34.39%.

The elevation and azimuthal radiation patterns of the optimal tapered CRF antenna, shown in Fig. (**5.5**), depict that the value of gain is 6.23 dBi. The value of directivity at f_r is found to be 7.63 dBi. The impedance plot, shown in Fig. (**5.6**), illustrates that at $f_r = 2.45$ GHz, the value of the real part of the impedance is 51.27 Ω which is nearly equal to the desired impedance of 50 Ω and the imaginary part is $- 0.18^0$, nearly equal to the desired 0^0. It means that good impedance matching has been achieved for the optimized antenna and it can be used for ISM band applications [30]. The performance of the designed optimal tapered CRF antenna is compared with the antenna of the rectenna system of Choi *et al.* [31] and the comparison is shown in Table **5.4**. This comparison shows that the proposed optimal tapered CRF antenna and the reference antenna have almost same f_r, S_{11}, and gain values.

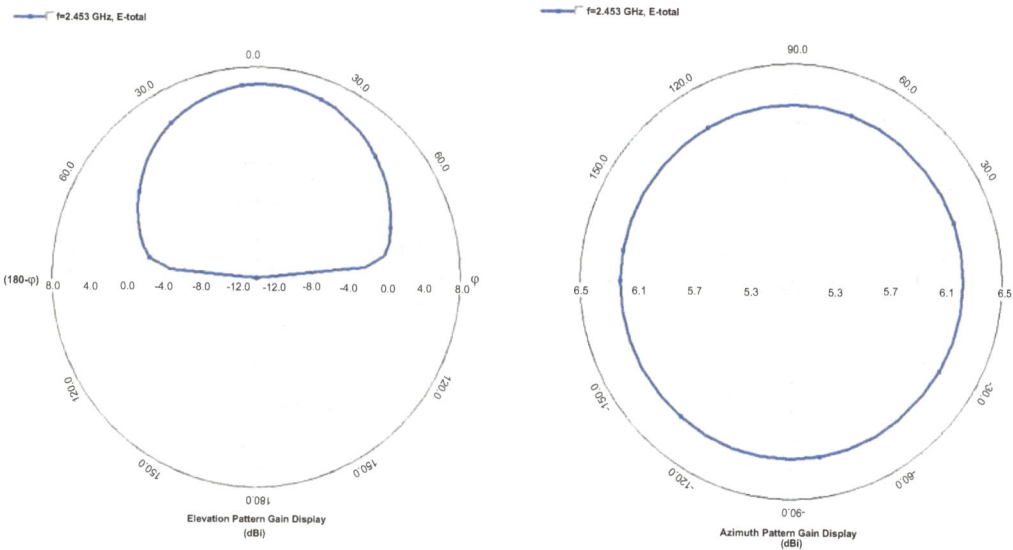

(a) **Elevation Radiation Pattern**   (b) **Azimuthal Radiation Pattern**

Fig. (5.5). Radiation Pattern Plots of Optimal Tapered CRF Antenna (Reprinted from the Springer Nature: Neural Computing and Applications, BFO-ANN Ensemble Hybrid Algorithm to Design Compact Fractal Antenna for Rectenna System, Dhaliwal, B.S. and Pattnaik, S.S. © 2016).

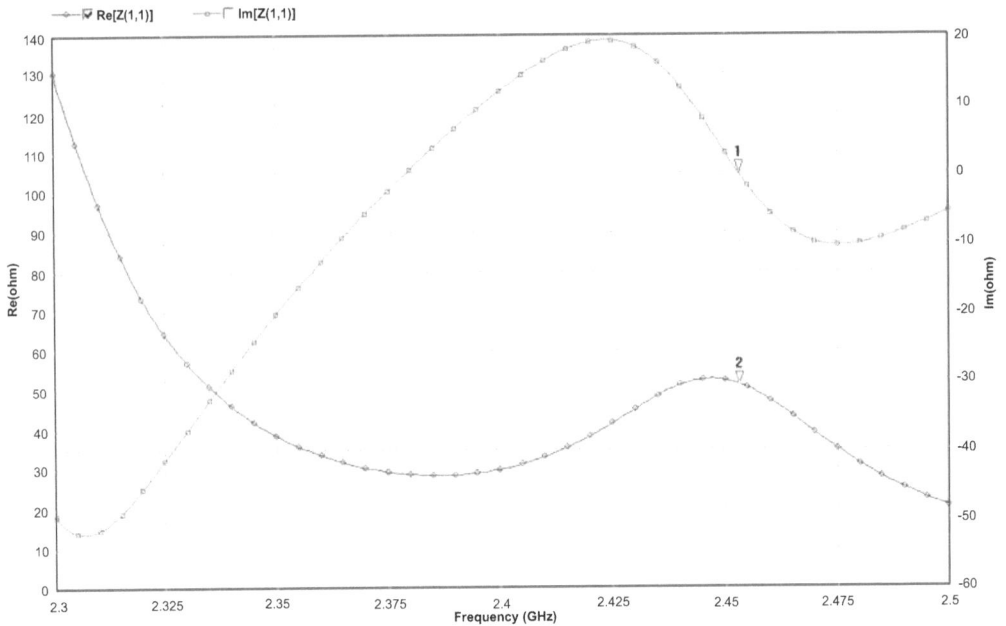

Fig. (5.6). Real and Imaginary Impedance Plot of Optimal Tapered CRF Antenna (Reprinted from the Springer Nature: Neural Computing and Applications, BFO-ANN Ensemble Hybrid Algorithm to Design Compact Fractal Antenna for Rectenna System, Dhaliwal, B.S. and Pattnaik, S.S. © 2016).

However, the area of the proposed optimal antenna is 65.61% relative to the area of a simple rectangular antenna for the same frequency as compared to 80% of that of the reference antenna. So, the proposed optimal tapered CRF antenna has less relative area resulting in a better size reduction capability. The impedance of the proposed tapered CRF antenna and the reference antenna is almost the same, so the proposed optimal antenna can be used in place of the antenna of Choi *et al.* [31]. Also, the bandwidth of the proposed optimal antenna is more than the reference antenna, so it will be matched for a wider band which will result in more received power and in turn, more output dc power leading to better conversion efficiency. Therefore, on the basis of the above discussion, the proposed optimal tapered CRF antenna is suggested as a better alternative to the antenna of Choi *et al.* [31].

Table 5.4. *: Comparison of Tapered CRF Antenna with Reference Antenna.

Antenna Parameter	Reference Antenna [31]	Optimal Tapered CRF Antenna
Resonant Frequency (GHz)	2.45	2.4531
Return Loss (dB)	42.55	37.24

(Table 5.4) cont.....

Antenna Parameter	Reference Antenna [31]	Optimal Tapered CRF Antenna
Gain (dBi)	7.58	6.23
Area (mm²)	1476.93	1248
Relative Area (%)	80	65.61
Impedance (Ω)	50.04	51.27
Impedance Bandwidth (MHz)	20	152

*Reprinted from the Springer Nature: Neural Computing and Applications, BFO-ANN Ensemble Hybrid Algorithm to Design Compact Fractal Antenna for Rectenna System, Dhaliwal, B.S. and Pattnaik, S.S. © 2016).

Experimental Results of Tapered CRF Antenna

The prototype of the CRF antenna having dimensions calculated as above is fabricated and is shown in Fig. (**5.7**). The antenna is fabricated using an RT-Duroid substrate of height 3.175 mm and fed using a co-axial SMA connecter.

Fig. (5.7). Prototype of Tapered CRF Antenna for ISM Band Applications (Reprinted from the Springer Nature: Neural Computing and Applications, BFO-ANN Ensemble Hybrid Algorithm to Design Compact Fractal Antenna for Rectenna System, Dhaliwal, B.S. and Pattnaik, S.S. © 2016).

The S_{11} results of the fabricated antenna are calculated using the Bird's site analyzer, and the measured results are shown in Fig. (**5.8**), which depict that the tapered CRF antenna resonates at a frequency of 2.502 GHz. The simulated value of S_{11} is 2.4531 GHz, so there is good matching between the experimental and simulation results. The slight shift in the resonant frequency is observed due to the fabrication error in feed-point, however, the bandwidth of the fabricated antenna is 152 MHz (6.19%), which is adequate to cover the whole 2.45 GHz ISM band [30].

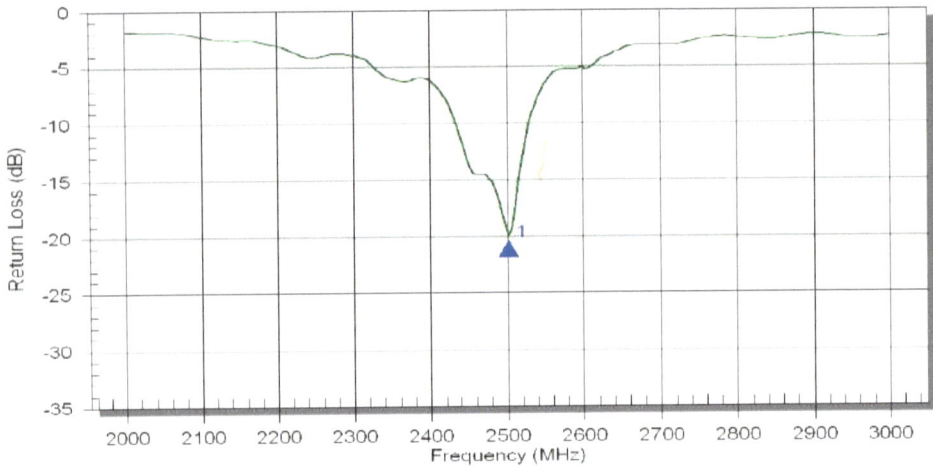

M1:(2502.11,-19.92)

Fig. (5.8). Measured S_{11} Results of Prototype of Tapered CRF Antenna (Reprinted from the Springer Nature: Neural Computing and Applications, BFO-ANN Ensemble Hybrid Algorithm to Design Compact Fractal Antenna for Rectenna System, Dhaliwal, B.S. and Pattnaik, S.S. © 2016).

The comparison of experimental and simulated azimuthal radiation patterns is shown in Fig. (**5.9**), which presents a good matching between the measured and simulated gain values.

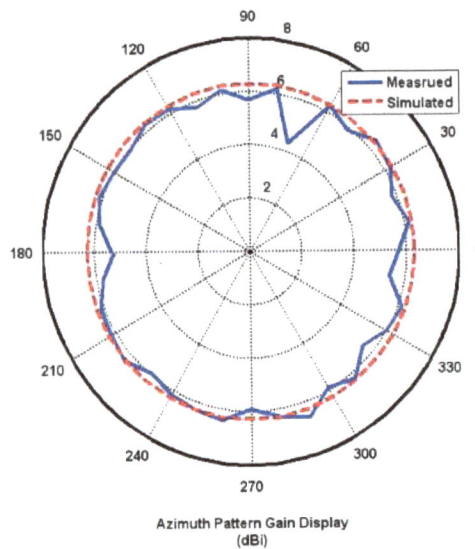

Azimuth Pattern Gain Display
(dBi)

Fig. (5.9). Simulated and Experimental Radiation Patterns of Tapered CRF Antenna (Reprinted from the Springer Nature: Neural Computing and Applications, BFO-ANN Ensemble Hybrid Algorithm to Design Compact Fractal Antenna for Rectenna System, Dhaliwal, B.S. and Pattnaik, S.S. © 2016).

PSO-ANN ENSEMBLE HYBRID MODEL TO DESIGN CCF ANTENNA FOR WLAN APPLICATIONS

The first iteration CCF antenna is discussed in Chapter 3, and it has been found that it has miniaturization characteristics. The ANN ensemble model developed for the estimation of f_r of CCF antenna for the given value of outer radius R is described in chapter 4. The design of CCF antenna for the desired user-defined frequency of operation involves the calculation of the required value of R, and then the other dimensions using the assumptions explained in chapter 3. In this section, the development of a hybrid PSO-ANN ensemble model to calculate the optimal radius R for the resonant frequency of 5.8 GHz is explained. In this model, the ANN ensemble model developed in chapter 4 has been used as an objective function of PSO algorithm to design a hybrid algorithm [32]. The flow chart of the proposed PSO-ANN ensemble hybrid algorithm is shown in Fig. **(5.10)**.

The error function given in equation (5.7) is used as the fitness function of the PSO algorithm, and it will converge to its minimum error value (ideally 0) for the optimal value of R of the CCF antenna resonating at 5.8 GHz. The f_r is the resonant frequency which is calculated using the trained ANN ensemble. The outer radius R of the CCF antenna is taken as the design variable.

$$Error = \sqrt[2]{(5.8 - f_r)^2} \qquad (5.7)$$

The initial parameters selected for PSO algorithm are: number of particles in the population = 30, Number of iterations = 100, and Parameters: $c_1 = c_2 = 2$. The optimal value of R of the CCF antenna provided by the algorithm is 7.9386 mm which is rounded off to two decimal places as 7.94 mm. Therefore, the CCF having the $R = 7.94$ mm should have a resonant frequency of 5.8 GHz.

To verify this, the first iteration geometry of the outer radius of 7.94 mm, the value provided by the hybrid PSO-ANN ensemble for 5.8 GHz frequency, is simulated using IE3D software and the S_{11} result obtained is shown in Fig. **(5.11)**, which shows that the optimal CCF antenna resonates at the frequency of 5.8 GHz. The feed point of the antenna has been finalized as (1.69, 4.95) by taking centre of the antenna at (0, 0) using the trial-and-error method. The matching of the simulation resonant frequency with the desired resonant frequency of 5.8 GHz validates the accuracy of the proposed hybrid PSO-ANN ensemble approach [32].

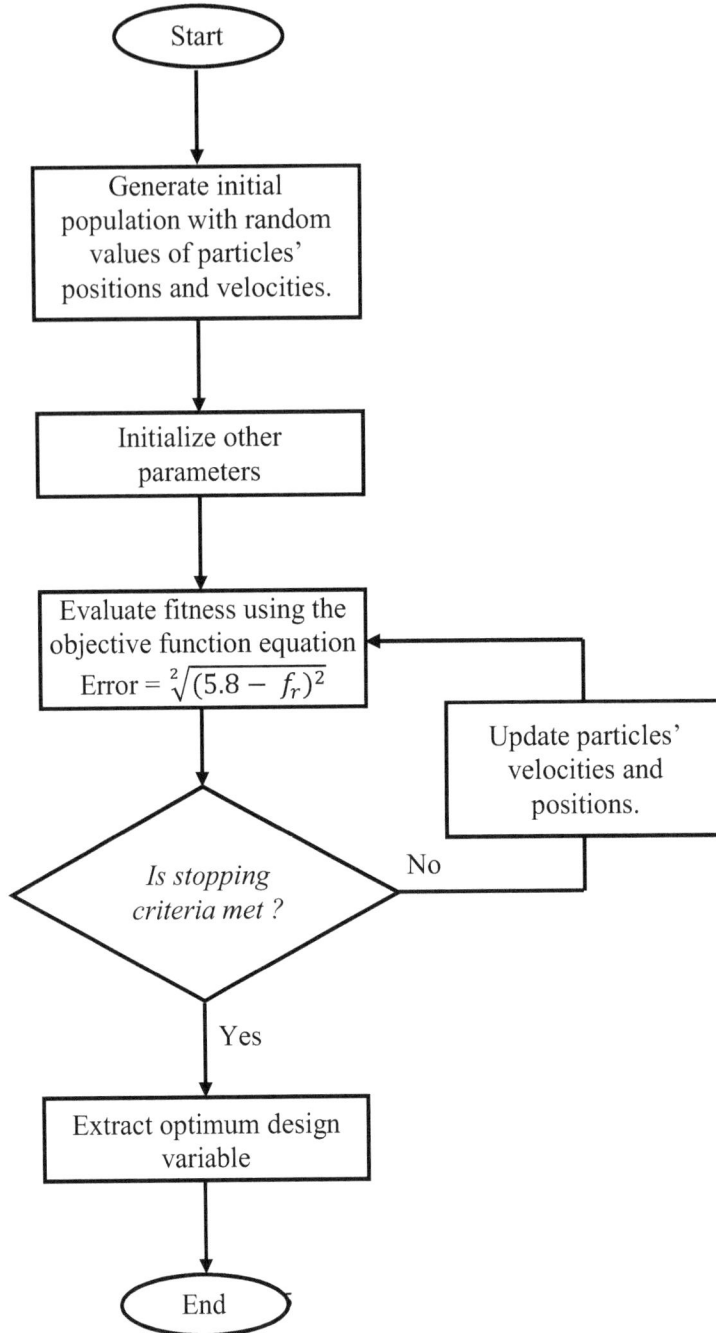

Fig. (5.10). Flow Chart of PSO-ANN Ensemble Hybrid Algorithm (Reprinted from the Springer Nature: Wireless Personal Communications, Development of PSO-ANN Ensemble Hybrid Algorithm and Its Application in Compact Crown Circular Fractal Patch Antenna Design, Dhaliwal, B.S. and Pattnaik, S.S. © 2017).

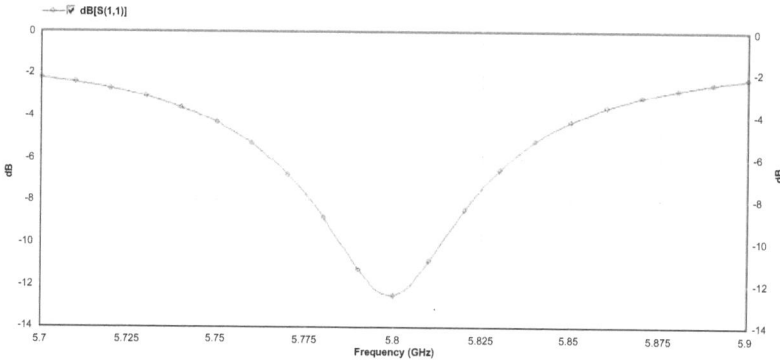

Fig. (5.11). S_{11} Plot of Designed CCF Antenna (Reprinted from the Springer Nature: Wireless Personal Communications, Development of PSO-ANN Ensemble Hybrid Algorithm and Its Application in Compact Crown Circular Fractal Patch Antenna Design, Dhaliwal, B.S. and Pattnaik, S.S. © 2017).

To calculate the miniaturization achieved, the radius of the simple circular microstrip patch antenna resonating at 5.8 GHz is calculated using the expressions given in equations 3.6 and 3.7, and then the area is calculated, which is compared with that of the designed antenna. The comparison shows that the CCF antenna has an area of 58.36% of the area of the simple circular microstrip antenna for the same frequency. Thus, the CCF antenna results in a size reduction of 41.64% at 5.8 GHz. The radiation patterns of the designed CCF antenna are shown in Fig. **(5.12)**, which depicts that the antenna has a peak gain of 6.16 dBi. So, the designed CCF antenna can be used for WLAN and other 5.8 GHz band applications.

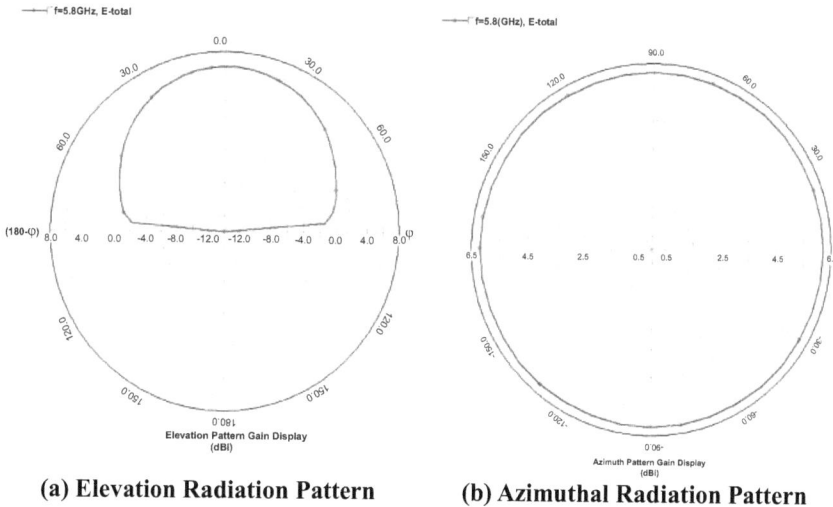

(a) Elevation Radiation Pattern   **(b) Azimuthal Radiation Pattern**

Fig. (5.12). Radiation Pattern Plots of Optimal CCF Antenna (Reprinted from the Springer Nature: Wireless Personal Communications, Development of PSO-ANN Ensemble Hybrid Algorithm and Its Application in Compact Crown Circular Fractal Patch Antenna Design, Dhaliwal, B.S. and Pattnaik, S.S. © 2017).

Experimental Results of CCF Antenna

A prototype of the CCF antenna having dimensions calculated using the PSO-ANN ensemble hybrid model is fabricated and is shown in Fig. (**5.13**). The return loss of the fabricated antenna is measured using Bird's site-analyzer for the 5.7 GHz to 6 GHz frequency range and is given in Fig. (**5.14**).

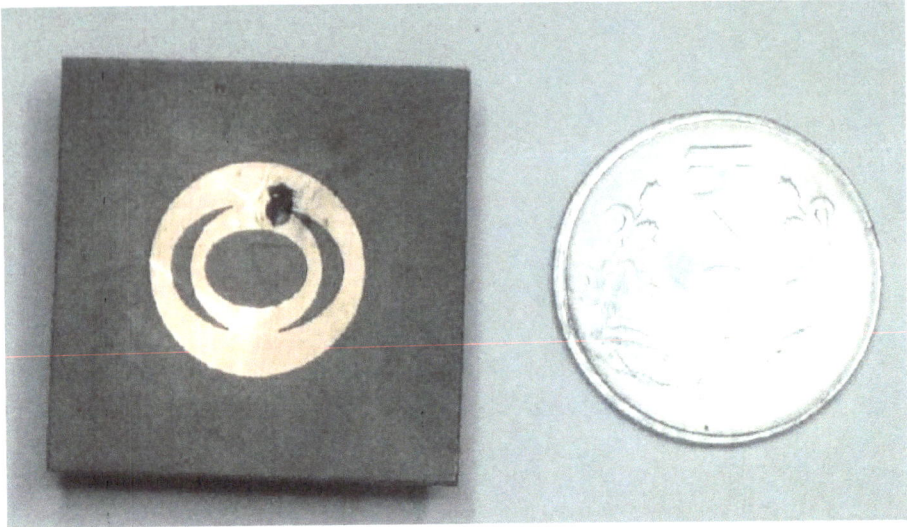

Fig. (5.13). Fabricated Prototype of CCF Antenna (Reprinted from the Springer Nature: Wireless Personal Communications, Development of PSO-ANN Ensemble Hybrid Algorithm and Its Application in Compact Crown Circular Fractal Patch Antenna Design, Dhaliwal, B.S. and Pattnaik, S.S. © 2017).

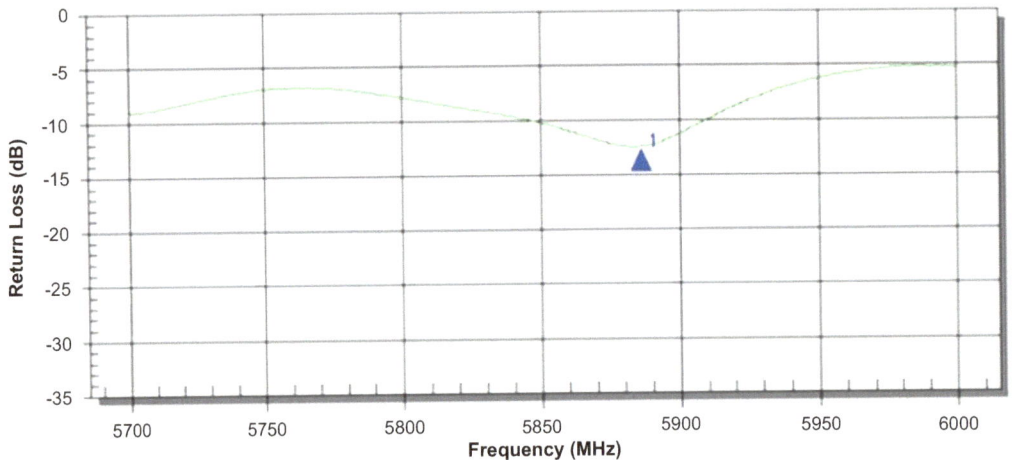

M1:(5886.08.-12.41)

Fig. (5.14). Measured S_{11} Results of CCF Antenna (Reprinted from the Springer Nature: Wireless Personal Communications, Development of PSO-ANN Ensemble Hybrid Algorithm and Its Application in Compact Crown Circular Fractal Patch Antenna Design, Dhaliwal, B.S. and Pattnaik, S.S. © 2017).

The measured results depict that the antenna resonates at the frequency of 5.886 GHz, thus having a good matching with the simulated f_r which is 5.8 GHz. So the proposed antenna can be used for WLAN applications in this band. The measured and simulated S_{11} results depict that the CCF antenna with this outer radius *i.e.,* 7.94 mm resonates at the desired frequency of 5.8 GHz, so, the accuracy of the PSO-ANN ensemble hybrid approach is also confirmed. The measured azimuthal radiation pattern is also in good agreement with the simulated radiation pattern, as shown in Fig. (**5.15**) [32].

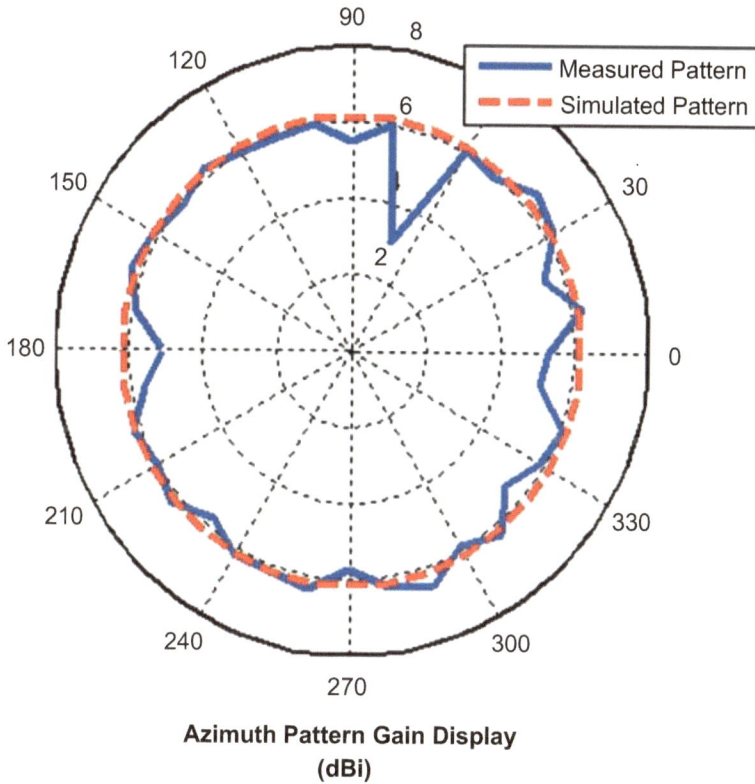

Azimuth Pattern Gain Display
(dBi)

Fig. (5.15). Simulated and Measured Azimuthal Radiation Pattern of CCF Antenna (Reprinted from the Springer Nature: Wireless Personal Communications, Development of PSO-ANN Ensemble Hybrid Algorithm and Its Application in Compact Crown Circular Fractal Patch Antenna Design, Dhaliwal, B.S. and Pattnaik, S.S. © 2017).

CONCLUSION

This chapter discusses the development of bio-inspired optimization techniques, namely GA, PSO, and BFO, for the design of fractal antennas for user-defined frequencies. The first application of these optimization techniques described in this chapter is for designing SGMF antenna. The formula found in the literature for calculating the resonant frequencies of this multiband antenna is simplified and used as an objective function of GA, PSO, and BFO algorithms. The dimensions of this antenna for desired resonant frequencies and for given substrate parameters are calculated using the proposed approach and results are compared to select the best optimization algorithm. The comparison on the basis of processing time and C_R shows that PSO is the best algorithm for this application.

The ANN models developed in the previous chapter are used as objective functions of optimization algorithms to design hybrid models for fractal antenna design. The development of these hybrid models is discussed in detail in this chapter. The first hybrid model is for designing the SGMF antenna. The MLPNN model developed for this antenna described in the previous chapter is used as an objective function of GA. The comparison of results shows that the hybrid model has improved performance as compared to the other design techniques.

This chapter also explains the use of the ANN ensemble model developed for tapered CRF antenna as an objective function of a BFO algorithm to estimate the dimensions of this antenna for ISM band application. The hybrid model results are compared with simulation results, and a good matching is obtained, which confirms the accuracy of the design method.

The next hybrid model explained in this chapter is the PSO-ANN ensemble hybrid model to design CCF antenna for WLAN applications. The ANN ensemble developed for this antenna in the previous chapter is used as an objective function of PSO algorithm for estimating the outer radius of the antenna for 5.8 GHz resonant frequency. The simulated results are obtained for the optimal antenna to evaluate the performance of the hybrid algorithm.

This chapter also presents the measured results of the proposed fractal antennas. The calibrated set-ups are used for experimentation. The measured results are compared with simulated results to validate the proposed designs. The experimental value of the resonant frequency of tapered CRF is 2.502 GHz and the simulated value of the same is 2.4531 GHz. The percentage difference in these two values is 1.95% which means a fine agreement of results is there. The percentage mismatch in the measured and simulated resonant frequency values of the RT-Duroid CCF antenna is 1.46% and that is 2.06% for FR4 CCF antenna. These values show that there is acceptable accuracy in the proposed design. The measured gain values are in tune with the simulated values, thus, validating our designed antenna geometries.

ANNEXURE 5.1: BFO-ANN ENSEMBLE HYBRID ALGORITHM CODING DESIGN STEPS

Part -1 ANN Ensemble coding Steps

1. Generate Primary Data Set using electromagnetic simulator like IE3D etc.

2. Generate required number of training and test data Subsets using Bagging Program (described in chapter 4) from primary data sets.

3. Initialize first ANN model as per specifications.

4. Train first ANN model using one of the subsets of data.

5. Test performance of trained model using corresponding test data subset.

6. Save trained ANN model if performance satisfactory otherwise repeat steps 4 and 5.

7. Repeat steps 3 to 6 to design desired number of ANN models.

8. Apply same input to all trained ANN models and get individual ANN model outputs.

9. Calculate average of the all individual ANN model outputs and it will be ANN ensemble output.

10. Check ANN ensemble performance by comparing the performance with individual ANN model outputs.

11. Plot the results if desired.

Part -2 BFO - ANN Ensemble Hybrid Algorithm Coding Steps

Initialize

Dimension of search space

Number of bacteria

Number of chemotactic steps

Limits the length of a swim

Number of reproduction steps

Number of elimination-dispersal events

Number of bacteria reproductions (splits) per generation

Probability that each bacterium will be eliminated/dispersed

Run length

Initial positions

Run following Loops for all bacteria

Elimination and dispersal loop

Reproduction loop

Swim/tumble(chemotaxis)loop

Report the results

Find min cost function for each bacterial using ANN Ensemble

Calculate the best value

Find optimal results

Repeat until improvement in results obtained

DISCLOSURE

Part of this article has previously been published in the following articles:

• B. S. Dhaliwal and S. S. Pattnaik, "Artificial neural network analysis of Sierpinski gasket fractal antenna: A low-cost alternative to experimentation," *Adv. Artif. Neural Syst.*, vol. 2013, pp. 1–7, 2013.

• B. S. Dhaliwal and S. S. Pattnaik, "Performance evaluation of Artificial Neural Networks in microstrip fractal antenna parameter estimation," in *Proceedings of 2012 IEEE International Conference on Communication Systems (ICCS)*, 2012, pp. 135-139.

• B. S. Dhaliwal and S. S. Pattnaik, "BFO–ANN ensemble hybrid algorithm to design a compact fractal antenna for rectenna system," *Neural Comput. Appl.*, vol. 28, no. S1, pp. 917–928, 2017.

• B. S. Dhaliwal and S. S. Pattnaik, "Development of PSO-ANN ensemble hybrid algorithm and its application in compact crown circular fractal patch antenna design," *Wirel. Pers. Commun.*, vol. 96, no. 1, pp. 135–152, 2017.

REFERENCES

[1]   R.W. Dearnley, and A.R.F. Barel, "A comparison of models to determine the resonant frequencies of a rectangular microstrip antenna", *IEEE Trans. Antenn. Propag.*, vol. 37, no. 1, pp. 114-118, 1989.
[http://dx.doi.org/10.1109/8.192173]

[2]   A. Hoorfar, "Evolutionary programming in electromagnetic optimization: A review", *IEEE Trans. Antenn. Propag.*, vol. 55, no. 3, pp. 523-537, 2007.
[http://dx.doi.org/10.1109/TAP.2007.891306]

[3]   T. Yu-Bo, Z. Su-Ling, and L. Jing-Yi, "Modeling resonant frequency of microstrip antenna based on neural network ensemble", *Int. J. Numer. Model.*, vol. 24, no. 1, pp. 78-88, 2011.
[http://dx.doi.org/10.1002/jnm.761]

[4]   S.S. Pattnaik, B. Khuntia, D.C. Panda, D.K. Neog, and S. Devi, "Calculation of optimized parameters of rectangular microstrip patch antenna using genetic algorithm", *Microw. Opt. Technol. Lett.*, vol. 37, no. 6, pp. 431-433, 2003.
[http://dx.doi.org/10.1002/mop.10940]

[5]   A.J. Kerkhoff, R.L. Rogers, and H. Ling, "Design and analysis of planar monopole antennas using a genetic algorithm approach", *IEEE Trans. Antenn. Propag.*, vol. 52, no. 10, pp. 2709-2718, 2004.
[http://dx.doi.org/10.1109/TAP.2004.834429]

[6]   N. Jin, and Y. Rahmat-Samii, "Hybrid real-binary particle swarm optimization (HPSO) in engineering electromagnetics", *IEEE Trans. Antenn. Propag.*, vol. 58, no. 12, pp. 3786-3794, 2010.
[http://dx.doi.org/10.1109/TAP.2010.2078477]

[7]   Y. Rahmat-Samii, J.M. Kovitz, and H. Rajagopalan, "Nature-inspired optimization techniques in communication antenna designs", *Proc. IEEE*, vol. 100, no. 7, pp. 2132-2144, 2012.
[http://dx.doi.org/10.1109/JPROC.2012.2188489]

[8]   S.V.R.S. Gollapudi, S.S. Pattnaik, O.P. Bajpai, S. Devi, C. Vidya Sagar, P.K. Pradyumna, and K.M. Bakwad, "Bacterial foraging optimization technique to calculate resonant frequency of rectangular microstrip antenna", *Int. J. RF Microw. Comput.-Aided Eng.*, vol. 18, no. 4, pp. 383-388, 2008.
[http://dx.doi.org/10.1002/mmce.20296]

[9]   N.D. Hieu, D.N. Chien, and N.K. Kiem, "Flexible PSO-based optimization of millimeter-wave triple-band antennas by the use of fractal configuration", *Proceedings of the 2010 International Conference on Advanced Technologies for Communications*, pp. 336-340, 2010.
[http://dx.doi.org/10.1109/ATC.2010.5672676]

[10]   A.P. Anuradha, A. Patnaik, and S.N. Sinha, "Design of custom-made fractal multi-band antennas using ANN-PSO", *IEEE Antennas Propag. Mag.*, vol. 53, no. 4, pp. 94-101, 2011.
[http://dx.doi.org/10.1109/MAP.2011.6097296]

[11]   D.H. Werner, P.L. Werner, and K.H. Church, "Genetically engineered multiband fractal antennas", *Electron. Lett.*, vol. 37, no. 19, pp. 1150-1151, 2001.
[http://dx.doi.org/10.1049/el:20010802]

[12]   M. Fernandez Pantoja, F. Garcia Ruiz, A. Rubio Bretones, R. Gomez Martin, J.M. Gonzalez-Arbesu, J. Romeu, and J.M. Rius, "GA design of wire pre-fractal antennas and comparison with other Euclidean geometries", *IEEE Antennas Wirel. Propag. Lett.*, vol. 2, pp. 238-241, 2003.
[http://dx.doi.org/10.1109/LAWP.2003.819694]

[13]   R. Ghatak, D.R. Poddar, and R.K. Mishra, "Design of Sierpinski gasket fractal microstrip antenna using real coded genetic algorithm", *IET Microw. Antennas Propag.*, vol. 3, no. 7, pp. 1133-1140, 2009.

[http://dx.doi.org/10.1049/iet-map.2008.0257]

[14]   M.F. Pantoja, F.G. Ruiz, A.R. Bretones, S.G. Garcia, R.G. Martín, J.M.G. Arbesu, J. Romeu, J.M. Rius, P.L. Werner, and D.H. Werner, "GA design of small thin-wire antennas: Comparison with Sierpinsky-type prefractal antennas", *IEEE Trans. Antenn. Propag.,* vol. 54, no. 6, pp. 1879-1882, 2006.
[http://dx.doi.org/10.1109/TAP.2006.875931]

[15]   R. Azaro, G. Boato, M. Donelli, G. Franceschini, A. Martini, and A. Massa, "Design of miniaturised ISM-band fractal antenna", *Electron. Lett.,* vol. 41, no. 14, p. 785, 2005.
[http://dx.doi.org/10.1049/el:20050774]

[16]   R. Azaro, E. Zeni, P. Rocca, and A. Massa, "Synthesis of a Galileo and wi-max three-band fractal-eroded patch antenna", *IEEE Antennas Wirel. Propag. Lett.,* vol. 6, pp. 510-514, 2007.
[http://dx.doi.org/10.1109/LAWP.2007.908009]

[17]   R. Azaro, L. Debiasi, E. Zeni, M. Benedetti, P. Rocca, and A. Massa, "A hybrid prefractal three-band antenna for multistandard mobile wireless applications", *IEEE Antennas Wirel. Propag. Lett.,* vol. 8, pp. 905-908, 2009.
[http://dx.doi.org/10.1109/LAWP.2009.2028627]

[18]   L. Lizzi, and A. Massa, "Dual-band printed fractal monopole antenna for LTE applications", *IEEE Antennas Wirel. Propag. Lett.,* vol. 10, pp. 760-763, 2011.
[http://dx.doi.org/10.1109/LAWP.2011.2163051]

[19]   W.C. Weng, and C.L. Hung, "An H-fractal antenna for multiband applications", *IEEE Antennas Wirel. Propag. Lett.,* vol. 13, pp. 1705-1708, 2014.
[http://dx.doi.org/10.1109/LAWP.2014.2351618]

[20]   B. Khuntia, S.S. Pattnaik, D.C. Panda, D.K. Neog, S. Devi, and M. Dutta, "Genetic algorithm with artificial neural networks as its fitness function to design rectangular microstrip antenna on thick substrate", *Microw. Opt. Technol. Lett.,* vol. 44, no. 2, pp. 144-146, 2005.
[http://dx.doi.org/10.1002/mop.20570]

[21]   S.A. Tretyakov, F. Mariotte, C.R. Simovski, T.G. Kharina, and J.P. Heliot, "Analytical antenna model for chiral scatterers: comparison with numerical and experimental data", *IEEE Trans. Antenn. Propag.,* vol. 44, no. 7, pp. 1006-1014, 1996.
[http://dx.doi.org/10.1109/8.504309]

[22]   D.W. Boeringer, and D.H. Werner, "Particle swarm optimization versus genetic algorithms for phased array synthesis", *IEEE Trans. Antenn. Propag.,* vol. 52, no. 3, pp. 771-779, 2004.
[http://dx.doi.org/10.1109/TAP.2004.825102]

[23]   M.A. Gondal, and A. Anees, "Analysis of optimized signal processing algorithms for smart antenna system", *Neural Comput. Appl.,* vol. 23, no. 3-4, pp. 1083-1087, 2013.
[http://dx.doi.org/10.1007/s00521-012-1035-x]

[24]   J.R. Perez, and J. Basterrechea, "Comparison of different heuristic optimization methods for near-field antenna measurements", *IEEE Trans. Antenn. Propag.,* vol. 55, no. 3, pp. 549-555, 2007.
[http://dx.doi.org/10.1109/TAP.2007.891508]

[25]   T. Datta, and I.S. Misra, "A comparative study of optimization techniques in adaptive antenna array processing: The bacteria-foraging algorithm and particle-swarm optimization", *IEEE Antennas Propag. Mag.,* vol. 51, no. 6, pp. 69-81, 2009.
[http://dx.doi.org/10.1109/MAP.2009.5433098]

[26]   M.A. Panduro, C.A. Brizuela, L.I. Balderas, and D.A. Acosta, "A comparison of genetic algorithms, particle swarm optimization and the differential evolution method for the design of scannable circular antenna arrays", *Prog. Electromagn. Res. B Pier B,* vol. 13, pp. 171-186, 2009.
[http://dx.doi.org/10.2528/PIERB09011308]

[27]   R. Mishra, R. Ghatak, and D. Poddar, "Design formula for Sierpinski gasket pre-fractal planar-

monopole antennas", *IEEE Antennas Propag. Mag.,* vol. 50, no. 3, pp. 104-107, 2008.
[http://dx.doi.org/10.1109/MAP.2008.4563575]

[28]   B.S. Dhaliwal, and S.S. Pattnaik, "Performance comparison of bio-inspired optimization algorithms for Sierpinski gasket fractal antenna design", *Neural Comput. Appl.,* vol. 27, no. 3, pp. 585-592, 2016.
[http://dx.doi.org/10.1007/s00521-015-1879-y]

[29]   C. Puente-Baliarda, J. Romeu, R. Pous, and A. Cardama, "On the behavior of the Sierpinski multiband fractal antenna", *IEEE Trans. Antenn. Propag.,* vol. 46, no. 4, pp. 517-524, 1998.
[http://dx.doi.org/10.1109/8.664115]

[30]   B.S. Dhaliwal, and S.S. Pattnaik, "BFO–ANN ensemble hybrid algorithm to design compact fractal antenna for rectenna system", *Neural Comput. Appl.,* vol. 28, no. S1, pp. 917-928, 2017.
[http://dx.doi.org/10.1007/s00521-016-2402-9]

[31]   D.Y. Choi, S. Shrestha, J.J. Park, and S.K. Noh, "Design and performance of an efficient rectenna incorporating a fractal structure", *Int. J. Commun. Syst.,* vol. 27, no. 4, pp. 661-679, 2014.
[http://dx.doi.org/10.1002/dac.2587]

[32]   B.S. Dhaliwal, and S.S. Pattnaik, "Development of PSO-ANN ensemble hybrid algorithm and its application in compact crown circular fractal patch antenna design", *Wirel. Pers. Commun.,* vol. 96, no. 1, pp. 135-152, 2017.
[http://dx.doi.org/10.1007/s11277-017-4157-8]

CHAPTER 6

Conclusion and Future Scope

Abstract: The conclusion drawn from the research work presented in the book, with some recommendations for future work, is presented in this chapter.

Keywords: ANN ensemble, Crown fractal antenna, Hybrid soft computing algorithm, Miniaturized antenna.

CONCLUSION

The presented research describes the development of fractal antennas suitable for low-power communication. The size reduction of the antennas is considered the main design parameter for selecting various geometries.

The fractal antennas developed to are: CRF antenna having size-reduction capabilities, tapered corner CRF antenna having enhanced bandwidth features in addition to size-reduction characteristics and CCF antenna having size-reduction characteristics. All the presented antennas have been analyzed on the basis of various antenna performance parameters like S_{11}, gain, radiation pattern plots, *etc.*, and acceptable performances are observed for desired applications.

The MLPNN, RBFNN and GRNN models have been developed for the estimation of various parameters of fractal antennas. The ANN model results are compared on the basis of different performance measures, and it is observed that the ANNs are very effective for fractal antenna analysis and design. The ANN ensemble models have resulted in further improved performances. The use of ANNs for fractal antennas has been found as an accurate and low-cost alternative to experimentation and lengthy simulation process. The ANN models have been developed to predict the required optimal dimensions of the fractal antenna for user-defined resonant frequencies, and to estimate other antenna parameters like S_{11}, gain, *etc.*

The GA, PSO and BFO-based algorithms are successfully employed for optimizing the fractal antennas to achieve desired specifications. The PSO algorithm has been found to be a better choice as compared to GA and BFO for designing an SGMF antenna. However, the GA-MLPNN hybrid model has further better performance than PSO for this design. A BFO-ANN ensemble hybrid model is designed to optimize tapered CRF antenna for ISM band applications. ANN ensemble models are employed as the objective function of GA and PSO algorithms to optimize the CCF antenna geometries for desired frequencies. It has been observed that the developed bio-inspired optimizing techniques are flexible, accurate and simple methods of designing fractal antennas for user-defined objectives.

The designed antennas are fabricated and tested with the calibrated measurement set-ups. The S_{11} results are obtained using the site-analyzer and the vector network analyzers. The gain measurements are also done for two fabricated antennas. The simulated results are compared with the experimental results, and a good matching of simulated and experimental results is seen. The proposed design approach based on ANN models, ANN ensemble models, bio-inspired optimization techniques and a hybrid of ANN and optimization techniques is validated by matching measured and simulated results.

So the presented research findings have contributed to the antenna design field by (i) new fractal antenna geometries having miniaturization and multiband features, (ii) ANN models to estimate fractal antennas parameters, (iii) ANN ensemble models for fractal antennas having improved performances as compared to single ANN models (iv) bio-inspired optimization algorithm based design approaches (v) hybrid approaches to design fractal antennas for desired applications and (vi) compact flexible fractal antenna geometry to meet the growing demand of wearable antennas.

Overall, this book has contributed to the development of new antennas for wireless applications and also a new method of antenna design and analysis for enhancing accuracy and efficiency.

FUTURE SCOPE

In the presented work, the antennas are optimized for a single optimization goal. In the future, the multi-objective optimization of the fractal antennas may be explored to optimize various antenna parameters simultaneously. The multi-objective optimization is useful, especially for designing multiband antennas. The design of the array of fractal antennas may be investigated in the future as the array results in enhanced performances compared to single antennas. Recently, the MIMO technology has attracted increased interest as it can improve the data transmission capacity and decrease multipath fading. So, the design of fractal antennas suitable for MIMO systems is another potential area for future work. The new fractal geometries, especially hybrid fractal geometries, *i.e.*, the fractal geometries designed by combining two or more fractal geometries, may be explored for fractal antenna design. The fractal geometries are expected to have better performance than the conventional fractal shapes but are relatively complex. So, the benefit of these geometries may be explored if high-end computing facilities are available. The latest optimization techniques, *e.g.*, SIMBO, having improved performances, may be explored for the design and analysis of fractal antennas. It is expected that these latest optimization techniques will use less computational resources and generate better results. Also, the design of new optimization techniques more suitable for fractal antenna design can be undertaken.

GLOSSARY

ANFIS	Adaptive Network-based Fuzzy Inference System
ANN	Artificial Neural Network
APE	Absolute Percentage Error
BFO	Bacterial Foraging Optimization
CCF	Crown Circular Fractal
CPW	Co-Planar Waveguide
CR	Coefficient of Correlation
CRF	Crown Rectangular Fractal
FCC	Federal Communications Commission
FDTD	Finite-Difference Time-Domain
GA	Genetic Algorithm
GPS	Global Positioning System
GRNN	General Regression Neural Networks
IFS	Iterated Function System
IMD	Implantable Medical Devices
ISM	Industrial, Scientific, and Medical
LTE	Long-Term Evolution
MAE	Mean Absolute Error
MAPE	Mean Absolute Percentage Error
MEMS	Micro-Electro Mechanical System
MICS	Medical Implant Communication Service
MIMO	Multi Input Multi Output
MLPNN	Multi-Layer Perceptron Neural Networks
MoM	Method of Moments
PSO	Particle Swarm Optimization
RBF	Radial Basis Function
RBFNN	Radial Basis Function Neural Networks
RF	Radio-Frequency
RFID	Radio Frequency Identification
SGMF	Sierpinski Gasket Monopole Fractal
SIMBO	Swine Influenza Model-based Optimization
UWB	Ultra-Wide Band

VHF   Very High Frequency

VSWR   Voltage Standing Wave Ratio

Wi-Fi   Wireless-Fidelity

WiMAX   Worldwide Interoperability for Microwave Access

WLAN   Wireless Local Area Network

WWAN   Wireless Wide Area Network

SUBJECT INDEX

A

Absolute percentage error (APE) 110, 111
Adaptive 41, 42, 54, 108, 110, 111
   antenna array processing 108
   network based fuzzy inference system
     (ANFIS) 41, 42, 54
Affine transformation 70
Algorithms 5, 37, 39, 41, 42, 43, 47, 48, 51,
    54, 95, 96, 100, 107, 109, 110
  beam-forming 5
  learning 41
  least-squares 42
  memetic 47
ANN 40, 42, 43, 44, 54, 85, 88, 135
   algorithms 44
   and optimization techniques 135
   applications 40, 43, 54, 85
   approach 44, 88
   methods 42
   principles 40
ANN ensemble 35, 42, 101, 129
   members 35
   method 42, 101
   output 129
Antenna(s) 3, 7, 11, 14, 15, 16, 18, 19, 29, 40,
    44, 49, 68, 70, 84
  designing compact microstrip 11
  enhanced bandwidth 14
  equilateral triangular microstrip 40, 84
  flexible 11, 16, 19
  monopolar 13
  reflector 44
  ring dielectric resonator 44
  ring monopole microstrip 49
  shape 15, 18
  single-band 7
  single microstrip 7
  small 70
  structures 68
  synthesis 29
  systems 3

  thin 44
Artificial neural networks 29, 89, 90, 91, 92,
    102, 130

B

Bandwidth 6, 7, 9, 10, 11, 13, 14, 15, 16, 18,
    19, 42, 43, 49, 76
  analysis 43
  broad 6, 7
  wide 8, 11
Behavior 6, 7, 12, 17, 28, 39, 70
  broadband 7
  dual band 6
  harmonic 70
BFO 39, 40, 107, 115, 116, 128, 135
  algorithm 39, 40, 107, 115, 116, 128
  based algorithms 135
Bio-inspired 28, 114
  algorithms 114
  computing 28

C

CCF antenna 77, 78, 80, 98, 99, 100, 101,
    102, 123, 125, 126, 127, 134, 135
  geometries 77, 78, 80, 135
  resonating 123
  results 125
Circuit 12, 15
  rectifier 15
Coefficients, hybridization 47
Communication devices 1, 2
Co-planar waveguide (CPW) 7, 9, 11, 13, 18,
    19, 20, 45, 52, 76
CRF antenna 72, 74, 75, 89, 90, 115, 119,
    121, 134
  designed 75
  designed optimal tapered 119
Crown rectangular fractal (CRF) 19, 72
CST software 14

www.ingramcontent.com/pod-product-compliance
Lightning Source LLC
Chambersburg PA
CBHW080021240326
41598CB00075B/622